D0402877

CHANCE OR PURPOSE?

CHRISTOPH CARDINAL SCHÖNBORN

CHANCE OR PURPOSE?

Creation, Evolution, and a Rational Faith

Edited by Hubert Philip Weber

Translated by Henry Taylor

IGNATIUS PRESS SAN FRANCISCO

Original German edition:
*Ziel oder Zufall?
Schöpfung und Evolution
aus der Sicht vernünftigen Glaubens*

Cover art:
Nautilus Shell © 2007 iStockphoto
and
Spiral Galaxy M51 photo © S. Beckwith (STScI)
Hubble Heritage Team, (STScI/AURA), ESA, NASA

Cover design by John Herreid

ISBN 978-1-58617-212-1
Library of Congress Control Number 2007928866
Printed in the United States of America ♾

The Christian idea of the world is that it originated in a very complicated process of evolution but that it nevertheless still comes in its depths from the Logos. It thus bears reason in itself.

—*Joseph Ratzinger*
Pope Benedict XVI

CONTENTS

FOREWORD TO THE GERMAN EDITION

Where do we come from? How did the world come into existence? These are fundamental questions that concern everyone. Those who hold the Christian faith, and theologians especially, have to make a serious attempt to explain what it means that we believe "in God, maker of heaven and earth". A series of scientific disciplines, such as biology and physics, are looking for answers to the question of how the world and man came into being. Are the answers of faith and those of science in competition with each other? Or can they exist independently of one another? Or is a co-existence even possible, in which each of the two approaches to reality retains its validity?

On July 7, 2005, an article by Cardinal Schönborn appeared in the *New York Times*, under the title of "Finding Design in Nature". In this article the Archbishop of Vienna took a critical look at some schools of thought whose evolutionary understanding of the world claims to "explain away" the Christian belief in creation.

This article brought reactions from many different sides, some of which were strongly polemical. There were two things Cardinal Schönborn did not intend, as he several times emphasized in public statements. On one hand, the valuable work done by many scientists engaged in honest research should not be belittled. On the other hand, "creationism"—that is, the view that the first chapter of the Bible should be understood literally as a report of events, and thus along the lines of a scientific text—is not an acceptable theological position. There is no bypassing an honest and serious discussion between natural science and theology, between knowledge and faith.

The discussion arising from this statement also brought positive results, and the dialogue between theology and science, for instance, was given a new impulse. One of the difficulties of dialogue is that on both sides there is often too little knowledge of the positions of the partner in the conversation, whoever that may be. Similar concepts are often used with different meanings. The conversation has to begin with the partners listening to each other, asking and answering questions and gaining a clear understanding of the limits of their own specialized knowledge so that a genuine dialogue may be initiated on that basis.

Cardinal Schönborn devoted the monthly catechetical lectures of the academic year 2005–2006 (lectures he holds on one Sunday evening each month in the cathedral of Saint Stephen in Vienna) to the theme of the theology of creation. The present book has grown out of those evening talks. The task of catechetics is to strengthen people's faith. That is why the Archbishop of Vienna, as a theologian, quite consciously presents the position of faith, and in doing so he goes into such questions as arise from natural science or are associated with it. It is not a matter of countering particular results or theories of natural science with other results and theories. We may assume rather, as the Second Vatican Council emphasized, that theology and natural science do not contradict one another,[1] since both are rational ways of approaching reality. A conflict may arise when one of them strays beyond its own sphere. In that sense, Cardinal Schönborn repeatedly distinguishes a scientific interest in the way that life evolved from an ideological view

[1] Cf. *Gaudium et Spes* 36:2.

that attempts to understand the world as a whole, starting from the theory of evolution. Cardinal Schönborn refers to this latter as "evolutionism" and consciously distances himself from it.

In nine stages, Cardinal Schönborn presents the Catholic belief in God the Creator and the Christian understanding of creation and of man as having been created by God.

The first chapter mentions the difficulties confronting any theology of creation today, particularly the relation between theology and science, between faith and knowledge. The decisive question turns out to be: Is it reasonable to talk about the world as a "creation" and to believe in a Creator? The Christian faith presupposes an affirmative answer to these questions, before all the other themes of the theology of creation.

The second chapter begins with the first verse of Holy Scripture, with the word of creation from the Book of Genesis. Following the text of the Bible, it starts with the question of what "creation" means at all, how we should understand the concept of "the beginning", and what Christian belief in the Creator means.

In the third chapter, the multiplicity of creation comes under scrutiny. The variety of species prompted the researches of Charles Darwin, just as it had done with many before him and continues to do with many after him. The Christian message insists that this variety is something intended by God.

The fourth chapter is devoted to an aspect that often receives too little consideration. Creation is not merely an act of God at the beginning of the world, but is continuing. Theology talks about continuing creation and about providence.

The subject of providence, however, also has another side to it: a critical challenge to our faith, which is articulated in the fifth chapter. If God guides everything, then how is it that there is so much suffering and injustice in the world?

The sixth chapter is about the creation of man and the question of whether, and in what sense, man can be considered as the "crown of creation". Are men a part of nature, or are they elevated above it? According to the theology of creation, both are true.

What does Jesus Christ have to do with creation? This is the subject of the seventh chapter. In the prologue to John's Gospel, we read, "All things were made through him" (Jn 1:3). Christ stands both at the beginning of creation and at its end, because we expect him to come again on the Last Day when the whole of creation will be completed and brought to perfection.

The eighth chapter draws practical conclusions from what has been said. What does God's commandment to man that he should have dominion over the earth actually mean? Where can we find the model that can offer points of reference for responsibility for creation, as properly understood?

The final chapter looks back again at the debate between theology and science, between the theology of creation and the theory of evolution, and attempts an interpretation. Current problems are addressed, and a suggestion is made of the lines along which the way these two spheres relate to each other might be understood.

— *The Editor, Verlag Herder*

I.

Creation and Evolution—The Current Debate

On the first page of the Bible we find, "In the beginning God created the heavens and the earth" (Gen 1:1). Believing in God the Creator, believing that he created heaven and earth, is the beginning of belief. That is how the Creed begins. That is the foundation on which everything else that Christians believe is based. Believing in God, and not believing that he is the Creator, would mean—as Thomas Aquinas once said—"not believing that God exists, at all". Belief in God as Creator is the foundation for all the other things we believe: that Jesus Christ is Savior, that there is a Holy Spirit, that there is a Church and an eternal life.

Creation: Where Does It Come from and Where Is It Going?

The *Catechism of the Catholic Church* emphasizes the fundamental significance of the belief in creation, when it says that this concerns the questions that every person asks himself, sooner or later in their lives, if he is leading a human life: "Where do I come from?" "Where am I going? "What is the aim of my life?" "What is its origin?" "What is its meaning?'.[1] Belief in creation also concerns the basis of ethics. For this implies that this Creator has something to say to us, through his creation, about the proper use of his work and the true meaning of our lives. Hence, since the time of the early Church, catechesis about creation has always been the foundation of all other catechesis.

If it is true that the question of origins ("Where do we come from?") is inseparable from the question of the end

[1] See CCC 282.

("Where are we going?"), then likewise the question of creation is always concerned with the question of the goal of things. It is thereby a matter of the question about a plan, about a "design" or "purpose". God did not just make his creation at one time, but he is sustaining it and guiding it toward a goal. This will be the subject of the fourth chapter, since this question is an essential part of basic Christian belief. God is not merely a Creator who once upon a time set everything going, like a clockmaker who has made a clock that will then run for ever more; rather, he is sustaining it and guiding it toward a goal. Creation, says Christian belief, is not just finished, but is *in statu viae*, on the way. As Creator, God also is involved in guiding and steering the world. We call this "providence" (even though the term is somewhat charged with historical associations). Christian belief insists that all of this—that is, that there is a Creator and a God who guides events—can also be recognized. Not of course in its entirety, in every detail, but fundamentally.

Do we know anything about this? A blind faith that simply demanded of us a leap into what was completely uncertain and unknown would not be a human faith. If belief in a Creator were completely devoid of all insight, with no way of knowing what believing in a Creator actually means, then such a belief would be inhuman. The Church has always quite rightly rejected that kind of "fideism", of blind faith.

Believing without knowledge, without the possibility of coming to know anything about the Creator, of our reason being able to comprehend anything about him, would not be Christian belief. The biblical and Judeo-Christian faith has always been convinced, not only that we can and should believe in a Creator, but also that we are able to understand a great deal about the Creator with our human reason.

In the Old Testament, in the Book of Wisdom (from the late second or early first century before Christ), there is a passage from which Paul quotes in the Letter to the Romans (1:19–20). It says there:

> For all men who were ignorant of God were foolish by nature;
> and they were unable from the good things that are seen to know him who exists,
> nor did they recognize the craftsman while paying heed to his works;
> but they supposed that either fire or wind or swift air,
> or the circle of the stars, or turbulent water,
> or the luminaries of heaven were the gods that rule the world.
> If through delight in the beauty of these things men assumed them to be gods,
> let them know how much better than these is their Lord,
> for the author of beauty created them.
> And if men were amazed at their power and working,
> let them perceive from them
> how much more powerful is he who formed them.
> For from the greatness and beauty of created things
> comes a corresponding perception of their Creator.
> Yet these men are little to be blamed,
> for perhaps they go astray
> while seeking God and desiring to find him.
> For as they live among his works they keep searching,
> and they trust in what they see, because the things that are seen are beautiful.
> Yet again, not even they are to be excused;
> for if they had the power to know so much
> that they could investigate the world,
> how did they fail to find sooner the Lord of these things? (Wis 13:1–9)

This classic passage is one of the bases for the conviction that was laid down as dogma—that is to say, as an explicit doctrine of the Church—by the First Vatican Council: that by the light of human reason we can come to know that there is a Creator who is guiding the world.[2]

The Bible accuses the pagans, who do not worship the true God, of idolizing nature and the world; of looking for mythical and magical forces behind nature and its phenomena. They make gods out of stars, out of fire, light and air. They have allowed themselves to be deceived. Their fascination with creation has misled them into idolizing created things. In that sense, the Bible is the first agent of enlightenment. In a certain sense, it "disenchants" the world; it divests it of its magical and mythical power, "demythologizes" the world, and "banishes the gods".

Are we aware that without this banishing of the gods from the world, even modern science would not have been possible? Only the belief that the world is created, that it is not divine, that it is finite, that it is "contingent", as we say in philosophical terminology, not "necessary"—that it might equally well not have existed—has made it possible for the world and everything that is in it to be studied for its own sake. We encounter finite, created realities and not gods or divine beings. This "disenchanting" of nature also has a painful side. Behind the tree, behind the spring, there are no longer hidden nymphs and divine beings, mythical and magical forces, but what the Creator put into them, something that human understanding can discover. That is why the Book of Wisdom says that God has created "all things by measure and number and weight" (Wis 11:20). That is the basis of all research into reality by the natural sciences.

Behind the things in the world there is the Creator's reason, which transcends all else. These things were made by

[2] Vatican Council I. *Dei filius*, chap. 2; Heinrich Denzinger, *Enchiridion symbolorum definitionum et declarationum de rebus fidei et morum: Kompendium der Glaubensbekenntnisse und kirchlichen Lehrentscheidungen*, ed. Peter Hünermann (Freiburg: Herder, 2005), 3004. Hereafter abbreviated DenzH.

him, and do not exist of themselves. They have been set free, so to speak, into their own existence. They are themselves, not of themselves, but because the Creator willed them to be, in a free and sovereign act of will. In that sense, they have their own autonomy, their own laws, their independence, their own being. A belief in creation makes it possible to see this.

While pagan antiquity, for the most part, "divinized" the world, idolized it, in the period of the rise of Christianity a philosophical reaction, the so-called "Gnosis" or "Gnosticism", devalued it. The world, and matter above all, are the result of an "accident", of a "fall" or deterioration. The material world is not actually something good, not something that was intended, supposed to be there, but is purely negative. Christianity rejected this view of Gnosticism, just as decidedly as it did the divinization of the world. Precisely because the world is created, the early Christians emphasized in the most decided fashion that matter was also created, that it is good, and that it bears meaning. It is not simply something "fallen", that broke away from a divine being that was originally one, and by an "accident within the divine being" arrived by "exclusion", so to speak, at this world of darkness. Matter is not something purely meaningless that we have to overcome. Matter has been created. "God saw that it was good" (Gen 1:10).

Man in this world of matter is not (as Gnosticism teaches) lost in a region of darkness, a divine spark that has fallen into the mire and that has to raise itself by returning to its divine origin. Man is part of this creation. He was intended by God as a microcosm that is an image of the macrocosm, as the being on the borderline who unites within himself both worlds, the spiritual and the material. "And God saw that it was very good", says the account of creation in the Book of Genesis after the creation of man (Gen 1:31). He belongs within creation and yet transcends it (see chapter 6).

Both the Gnostic and the divinizing view are irreconcilable with the biblical belief in creation. The biggest stumbling block for the people in the ancient world, however, was certainly the belief that God creates out of nothing, with no preconditions, *ex nihilo*. I think this is still the key question today in the whole debate surrounding creation and evolution. What does it mean to say that God creates? The great difficulty we have—on which Charles Darwin likewise came to grief—is that we have no concept of what it means that God is Creator, no way of visualizing it, no way of conceiving it. For everything we know consists merely of changes. Those who created the cathedral of Saint Stephen in Vienna did not create it from nothing. They marvelously shaped wood and stone. All the myths and epics of creation outside the Bible start by supposing that a divine being formed this world from what was there already. Creation out of nothing, the absolute and sovereign act of creation, as recounted in the Bible, is unique. We shall see how this is of fundamental importance for an understanding of creation as having an independence intended by God.

Theology of Creation and the Worldview of Natural Science

Belief in creation acted as godfather by the cradle of modern science. (I am unable to demonstrate this in detail here.) Nicholas Copernicus, Galileo Galilei, and Isaac Newton were convinced that science is concerned with reading what is written in the book of creation. God wrote this book, and he gave man understanding so that he could make out what it says. God wrote the book in a legible script, which is certainly not easy to make out, but it can be read. All the work of science lies in discovering order, laws, and connections. Let us express this with the metaphor of a book: it is discovering the alphabet, the grammar and syntax, and finally the text that God has written in this book of creation.

It is one of the more tenacious "myths" of our epoch—indeed, I would say, one of the well-established prejudices—that relations between science and the Church are bad, and that faith and science exist, from ages past, in a kind of persistent conflict. The Church is regarded as the great brake on progress, and science as the courageous liberator. Above all, the "case of Galileo", in the popular version, is depicted in that way, with the researcher as the victim of the dark Inquisition. A great deal of that, however, is *legenda nigra*, a "black legend", drawn in sharp contrasts during the Enlightenment, but not entirely doing justice to the historical truth. Other examples could be cited in which belief in creation was the rational basis for scientific research. Thus, for instance, Gregor Mendel, an Augustinian friar and scholar from Brünn who worked out a theory of inherited characteristics, is one among many people without whose researches we cannot nowadays imagine the world.

Belief in God as Creator is not an obstacle but rather the opposite. Why should the belief that the universe has a Creator stand in the way of science? Why should it in any way cause problems for science, if scientists understand their research, their discoveries and the theories they evolve, their comprehension of relationships, as "studying the book of creation"? There are in fact an enormous number of scientists who bear witness to this, who not only make no secret of their faith, but positively affirm it and see no conflict between science and their faith. The fact that nonetheless, there repeatedly have been conflicts and perhaps still are, is something we would have to look at separately in each case.

Two examples will demonstrate the Church's fundamental conviction that faith and science do not, in principle, conflict with each other. In 1870, the First Vatican Council declared:

> Though faith is above reason, there can never be any real discrepancy between faith and reason. Since the same God who reveals mysteries and infuses faith has bestowed the light of reason on the human mind, God cannot deny himself, nor can truth ever contradict truth.[3]

The conclusion to be drawn from this is that neither the Church nor science should be afraid of the truth. It makes us free, says Jesus (see Jn 8:32). The second quotation is from the Second Vatican Council. The Council's Constitution *Gaudium et spes*, which concentrates more on the question of "natural science and faith", makes the following statement:

> Therefore if methodical investigation within every branch of learning is carried out in a genuinely scientific manner and in accord with moral norms, it never truly conflicts with faith, for earthly matters and the concerns of faith derive from the same God. Indeed whoever labors to penetrate the secrets of reality with a humble and steady mind, even though he is unaware of the fact, is nevertheless being led by the hand of God, who holds all things in existence, and gives them their identity.[4]

Why do conflicts arise, then, time after time—or at least (following my article in the *New York Times* of July 7, 2005, for instance), violent polemics? Indeed, these latter may prove to be quite positive in furthering the discussion.

Conflicts may arise from misunderstandings. It may happen that we do not express ourselves with sufficient clarity,

[3] *Dei filius*, chap. 4; DenzH 3017; CCC 159.

[4] *Gaudium et spes* 36:2; also quoted in a different translation in CCC 159.

or even that our ideas may not be clear enough. Misunderstandings of this kind can be cleared up. I have already mentioned one of the most frequent misunderstandings, namely, the misunderstanding about the Creator himself. We will show this in the next chapter, taking the work of Charles Darwin as an example, when it is a matter of how the Creator acts. There seems to me no danger—nowadays, at least—of the Church's trying to speak on behalf of science. Yet we do repeatedly encounter the difficulty of the boundaries on each side not being mutually recognized and respected. That is why they have to be sounded out and talked through, time and again.

The grandiose successes of the natural sciences have, time and again, tempted people to stray beyond their boundaries. The impression is given that religion is constantly retreating before the tremendous progress of science, that it has had to give up more and more fields, because more and more has been explained by science. More and more realms that were supposedly explained previously in a "primitive supernatural way" can now be given a "natural" explanation—and that usually means attributing to them purely material causes. When Napoleon asked Laplace whether there was any place for God in his theories, the latter is supposed to have replied, "Sir, I do not need that hypothesis." God appears as a redundant hypothesis, as a "crutch for cripples" or for people who cannot yet stand on their own two feet. Man is achieving more and more freedom from his old dependencies. He is also emancipating himself from God as an explanation, and perhaps he does not need him at all any more.

When in 1859 Darwin published his famous and historically highly influential book *On the Origin of Species*, the basic message was that he had discovered a mechanism that explained a self-operating mode of development for plants and animals, without requiring a Creator. As he himself says, it was a matter of discovering a theory which, for the development of species from lower to higher, did not require new and ever more perfect kinds of creation, but simply made do with chance variations and the survival of the fittest—an explanation that can do without creation.

There is no doubt that Darwin's principal work was a stroke of genius, and it remains one of the truly great works in the history of ideas. With an incredible gift for observation, and a great deal of hard work and prodigious mental powers, he produced this seminal book, which is among the most influential works in the history of ideas. And he could see in advance that many areas of science would benefit from his research. We can in fact say, today, that the model of "evolution" has become a universal key to understanding, and its use has spread to many areas of knowledge.

Its success is certainly not to be attributed purely to scientific factors. Darwin himself and, above all, his advocates, who created "Darwinism", so to speak, gave his theory a strong general philosophic character. Many saw the *Origin of Species* as an alternative to "the view", as Darwin said, "that the individual species were independently created". Individual acts of creation were no longer necessary in order to explain the origin of all the species. The famous final sentence of the second edition of the *Origin of Species* certainly does still leave a place for the Creator, but it is greatly reduced:

> There is grandeur in this view of life, with its several powers, having been originally breathed by the Creator into a few forms or into one; and that, whilst this planet has gone cycling on according to the fixed law of gravity, from so simple a beginning

> endless forms most beautiful and most wonderful have been, and are being, evolved.[5]

I believe that this statement was genuinely intended to be pious. Yet behind it there is a concept of creation that in theology is called "Deism". According to this view, an act of creation stands at the beginning of everything. God infused into a single form the seed of all life. From that beginning, it developed according to laws. And Darwin attempted to discover these, to describe them and formulate them. There is no need for any further interventions by God here. But is this concept of creation appropriate? Does "creation" mean that God intervenes here and there? What does "creation" actually mean? These questions will need to be clarified.

The world has been keenly involved in the philosophical controversy concerning Darwin's theory, "Darwinism", for almost 150 years. I will mention just three examples of interpretations with a clearly philosophical slant. In 1959, in a speech to mark the one hundredth birthday of Darwin's famous book, Julian Huxley said:

> In evolutionary thought, there is no need or room for the supernatural, and no longer any place for it. The earth was not created, it evolved.... So did all the animals and plants that inhabit it, including our human selves, mind and soul as well as brain and body. Religion, too, originated by evolution.... The evolutionary man can no longer take refuge in the arms of a divinized father figure whom he has himself created.

In my opinion, this is not a scientific statement but a philosophical one, the expression of a way of looking at the

[5] Charles Darwin, *On the Origin of Species by Means of Natural Selection, or the Preservation of Favoured Races in the Struggle for Life*, 2nd ed., 1859; (London: J.M. Dent and Sons, Ltd., 1971), p. 463.

world. Basically, it is the "confession" of a belief in materialism. Thirty years later, in 1988, the English author Will Provine wrote the following in an essay on evolution and ethics:

> Modern science directly implies that the world is organized according to strictly mechanical principles. There are in nature no principles whatever directed towards goals. There are no gods, and no powers can be rationally ascertained which devise or plan anything.

Four years later, Peter Atkins, a professor of chemistry at Oxford, made a similar statement:

> Humanity should accept that science has swept away any justification for belief in the universe having a meaning or purpose, and the fact that the belief in a purpose has survived is thanks only to feelings.

Those are not actually scientific statements but philosophical ones. They might also be regarded, to some extent, as confessions of belief. Such statements, and similar ones, may be heard time and again, and they are the reason that I said in the *New York Times*, regarding this kind of straying beyond the boundaries of competence (but not about scientific theories as such), that this is not science but ideology, a way of looking at the world.

In the Bible's Book of Wisdom, the words are put into the mouths of those who deny God, "We were born by mere chance, and hereafter we shall be as though we had never been; because the breath in our nostrils is smoke, and reason is a spark kindled by the beating of our hearts" (Wis 2:2). We might almost describe these lives as a materialist's confession of belief: Even my mind is just a product of matter.

Not Losing Wonder

What prevents man from recognizing the Creator? What prevents his drawing the conclusion, from the sublime beauty of creatures, that there is a Creator? Today, two thousand years later, such a conclusion actually ought to be much easier to draw, since we know so incomparably much more than then. Who could then suspect the immeasurable vastness of the universe? It does indeed say, in the Bible, "numerous as the stars in the heaven and as the sand which is on the seashore" (Gen 22:17), yet could people know, at that time, that the number of stars is indeed comparable to the grains of sand on the seashore? There are so many suns there are in this universe! Could anyone suspect, then, how incredibly complex an atom is? Could people be aware, then, how incredibly fascinating it is, the way a single cell functions? Has this increase in knowledge somehow forced us to give up believing in a Creator? Has this knowledge somehow ousted the Creator? Or does it not, on the contrary, make much better sense now, is it not much more reasonable to believe in a Creator? May not belief in a Creator actually have become much easier?

Perhaps it is actually only the notion that a Creator somehow intervenes in this wondrous work of nature that is—quite rightly—being rejected. Perhaps it also arises from the fact that our understanding based on faith has not kept up with scientific knowledge, that we still have "the faith of a child" alongside an incredibly developed scientific knowledge. In this sense, I am happy that my article has sparked off such a debate, which will perhaps also lead people to reflect further upon the questions of "creation and evolution" and of "faith and science".

I see no difficulty in combining faith in a Creator with the theory of evolution, subject to one condition: that the three boundaries of scientific theory are adhered to. There is no question that the three passages I have just quoted cross that boundary. If science sticks to its own methods it cannot come into conflict with faith. And yet perhaps it is difficult to keep within those boundaries, since we are not just scientists but also men with feelings, men who are struggling for faith, men looking for the meaning of life, who hence—as scientists—unavoidably bring with us questions about the way we see the world.

In 1985, a symposium was held in Rome under the title "Christian Faith and the Theory of Evolution", at which I was privileged to read a paper. Cardinal Ratzinger, now Pope Benedict XVI, presided at this symposium, and Pope John Paul II received us in audience at the close of the symposium. He said then,

> A belief in creation, rightly understood, and a rightly understood doctrine of evolution, do not stand in each other's way. Evolution presupposes creation; creation, seen in the light of evolution, appears as an event extended over time—as a *creatio continua*, as a continuing creation—in that God becomes visible, to the eye of faith, as the 'creator of heaven and earth'.

Pope John Paul II added, however, that in order for belief in creation and the doctrine of evolution to be correctly understood, this requires the intervention of reason, of philosophy, of reflection. For me, this question is not primarily one of faith and knowledge but a question of reason. It is entirely rational to assume that there is a significance in the development of nature, even though the methods of natural science quite rightly require this to be set in brackets. Yet my common sense cannot be excluded by scientific method. Reason tells me that there is order and a plan, meaning and purpose, that a clock has not come into being by chance, and far less still the living organism of a plant, an animal, or indeed a human being. That is why we should wonder, for wonder is the beginning of philosophy.

I am grateful for the immense achievement of the natural sciences. They have brought an incredible increase in our knowledge. They do not diminish belief in creation—on the contrary, they strengthen me in believing in the Creator, and believing in how wisely, how wonderfully he has created everything.

II.

"In the Beginning, God Created ..." (Genesis 1:1)

The March of the Penguins is said to be a beautiful film, and in just a few weeks it became a world-wide success. The life of these "waddling birds" and the way they behave as a flock is fascinatingly portrayed, as is the way they overcome extreme climatic conditions. Yet directly after its release, there followed another dispute about evolution. American Christian commentators were enthusiastic about the penguins' virtues, and held that their ability to withstand the extreme temperatures, to survive both the sea and predators, and at the same time to be examplary parents, self-sacrificing and monogamous, argues against Darwin's theory and in favor of an "intelligent design"—that is, for creation and against Darwin. The director of the film, a French film-maker, defended himself energetically against being co-opted in this way. He was "brought up on the milk of Darwin", so he said, and had "only" wanted to make a film about animals.[1]

I think that controversy is typical today. The atmosphere is heated, and both sides are ready with accusations. The controversies are almost reminiscent of something like the *Kulturkampf*, the clash of cultures. For example, Salman Rushdie writes in the *New York Times*, and also in *Die Zeit*, attacking religions in the most violent way—no peace is possible with them, in his view. Thus he says, "Throughout the world, Islamic voices are declaring that the doctrine of evolution cannot be reconciled with Islam." The theory of "intelligent design" is for him "the theory that tries retrospectively to impose the antiquated concept of a Creator on the beauty of creation". His view is that "a certain crudity is appropriate" in such a context.

[1] See *Der Standard*, October 22–23, 2005, p. 43.

Similarly, a great deal of aggressive polemic was directed against those "who assert that God created them", as we could recently read in *Die Zeit*.[2] People who assert these things are called "fanatics"—and perhaps there really are a few among them, or some of them behave like fanatics; but the fact that someone believes God has created them still does not justify running down such a belief in that way. In the time of Darwin, according to the writer in *Die Zeit* we have just quoted, "most people" subscribed to "crude religious creation-myths", and this is obviously long outdated today. Quite unpolemically, one might pose the counter-question whether people who are enthusiastic about *The Creation*, that marvelous oratorio by Joseph Haydn, are subscribing to "crude myths"?

Belief in Creation as Fanaticism?

Is everyone who believes "a God created them" just a blind fanatic? Or is our deep pleasure in Haydn's *The Creation* just a romantic surge of the spirit? Can a rational person believe in a Creator at all? On this point, I just want to listen, without polemics, to what faith and reason say about it. In reaction to my article in the *New York Times*, a scientist wrote to me that he would like to believe, but he simply could not "believe in a Creator God, an old man with a long white beard". I replied to him that no one actually expected him to believe such a thing. On the contrary, that kind of conception of the Creator—perhaps childlike, but certainly childish—is a long way from what the Bible says

[2] Urs Willmann, "Immer Ärger mit den Verwandten", in *Die Zeit*, September 29, 2005.

about the Creator, and what the clause in the Creed "I believe in God, the Father, maker of heaven and earth", means. In my letter to the scientist, I responded that it would be a good thing if his scientific knowledge and his religious knowledge were a little more nearly on the same level, and if his high level of knowledge as a scientist were not accompanied by a religious knowledge still that of a child.

At this point we should also mention another frequent misunderstanding. It concerns so-called "creationism". Often nowadays in polemics, belief in creation is lumped together with "creationism". Yet believing in God the Creator is not identical with the way that, in some Christian circles, people try to understand the six days of creation spoken of in the first chapter of the Book of Genesis as if this had been literally reported, as six chronological days, and try by all possible arguments, even scientific ones, to prove that the earth is about six thousand years old. Attempts like that to take the Bible literally, as if it were making scientific statements at this point, are what is called "fundamentalism". To be more exact, in American Protestantism this view of the Christian faith has called itself "fundamentalism" from the start. Starting from a belief that every word of the Bible was directly inspired by God—that is, starting from an understanding of literal inspiration—the six days of the creation are also taken to mean what they say, word for word. It is understandable that many people in the U.S.A. are energetically opposed to this view—even so far as going to court and taking legal action against such things being taught in schools. There is, of course, also the legitimate concern with critical questions about teaching "Darwinism"—but that is a different matter.

The Catholic position on "creationism" is clear. Saint Thomas Aquinas says that one should "not try to defend the Christian faith with arguments that make it ridiculous, because they are in obvious contradiction with reason".[3] It is nonsense to maintain that the world is only six thousand years old. An attempt to prove such a notion scientifically means provoking what Saint Thomas calls the *irrisio infidelium*, the mockery of unbelievers. Exposing the faith to mockery with false arguments of this kind is not right; indeed, it is explicitly to be rejected. Let that be enough on the subject of "creationism" and "fundamentalism".

An Outline of the Theology of Creation

What does the Christian faith say about belief in God the Creator and creation? The classic Catholic doctrine is expounded in the *Catechism of the Catholic Church* (CCC) and now in an abbreviated form in the *Compendium* of the *Catechism*. It has four basic elements:

1. The doctrine of creation implies that there is an absolute beginning—"In the beginning, God created heaven and earth"—and that this absolute beginning was the free and sovereign constitution of being out of nothing.
2. Part of the Christian doctrine of creation holds that creatures are different from each other. God created "all kinds" of animals, as it says in Genesis (1:21, 25). The great variety of creatures is the subject of our next chapter.
3. We believe, not just in an absolute beginning, but also that creation is being sustained. God keeps in being everything that he created. That is his continuing work

[3] *Summa Theologiae* I, q. 32, a. 1 resp.

of creation, which theology calls "*creatio continua*", "continuing creation".

4. Part of what creation also means is that it is guided. God did not just start creation going once, at the beginning, and since then he is letting it run; divine providence is a part of the doctrine of creation. God is guiding his work to a goal (see below, chapter 5).

We have thus sketched out in advance the substance of the rest of our program. Yet this is not *merely* a matter of Christian doctrine; rather, in dealing with each theme my aim is to seek a discussion with the natural sciences, so far as this is possible for someone who is a layman in scientific matters. In particular, we are concerned time and again with the question of the relation between belief in creation and the theory of evolution.

What Is the Beginning?

Let us start with the question of the absolute beginning. Nowadays the theory of the original explosion, the Big Bang, is generally admitted as a scientific theory about the beginning of the universe. Seventy-five years ago, the American astronomer Edwin Hubble discovered that our universe is expanding with inconceivable rapidity.

The universe, therefore, once must have begun to expand—with a Big Bang, in fact—from being highly concentrated, exceedingly dense at the beginning. It began to expand "like an explosion", so to speak. This theory is strengthened by various observations, above all by the so-called "cosmic background noise", which is interpreted as a dying echo of the Big Bang. In any case, many questions remain a mystery, and probably cannot even be answered by the theory itself, but are a challenge to the researchers' rational powers.

First, there is the simple question: Where is the universe expanding into? Is it expanding into a space? Yet there is no space outside of the universe, the cosmos, with its gigantic dimensions of fourteen billion light-years—such is the general assumption; a light-second corresponds to perhaps three hundred thousand kilometers (around one hundred and eighty-nine thousand miles). According to the most recent research, it could be as high as forty-six billion light-years, but with these almost inconceivable numbers and dimensions, that hardly matters any more. Our own galaxy, our 'Milky Way' alone, is one hundred thousand light-years in size. Who can visualize that? Beyond this gigantic dimension of the cosmos, there is no space. The space in which we are living "originated with the Big Bang, and since then has been expanding".[4]

The question of time is just as great a riddle. The Big Bang does mean that the universe has a beginning and is moving toward an end. Now, there is a great temptation to ask, "What was there before the beginning?"—The answer can only be, "Just as space only exists because of the expansion of the universe, and in the place where it is expanding, so it is with time. There is no time before time. Time and space originate with the Big Bang." Only within the cosmos, and with it, is there time.

In recent decades, scientists have tried to get back closer and closer to the Big Bang. Nobel prize-winner Steven Weinberg wrote a famous book entitled *The First Three Minutes*.[5] It deals with the way the universe came into being. It is fascinating to find out what researchers today say about the

[4] *Spektrum der Wissenschaft*, May 2005, p. 41.

[5] Stephen Weinberg, *The First Three Minutes* (New York, 1977).

first decisive moments following the Big Bang. Everything that developed later—galaxies, stars, planets, life on this earth—all of this was decided in the very first moments. The well-known physicist Professor Walter Thirring writes, "If the Big Bang had been too feeble, and the whole thing had collapsed again, then we would not exist. If it had been too violent, then everything would disperse too quickly",[6] and again, we would not exist. He compares the beginning of the world to launching a rocket so as to put a satellite into orbit around the earth. He says, "If you use too little fuel, it falls back down. If you use too much, it shoots off into space." But he adds that in the Big Bang, the conditions for "accuracy" in the first few moments were incomparably tighter than in the launching of a rocket to put a satellite into orbit. It is so far "beyond human capacity to conceive" the degree of precision in this event, which occurred in microscopic fractions of the very first seconds, so Professor Thirring says, that he exclaims, "And that is supposed to have happened by chance—what an absurd notion!"[7]

Have we thereby reached the point at which belief in creation intervenes? Is the threshold, beyond which the Creator is standing, here at the frontier of what science has achieved? Careful! The idea that God "made" the Big Bang, that in these microscopic fractions of the first seconds we have, so to speak, run up against the last barrier, beyond which only the Creator can explain how that happened—it is easy to think so. Many scientific discussions, and even theological discussions, are haunted by this notion. Did God, so to speak, "blow the whistle" to start the great game of the universe?

[6] Walter Thirring, *Cosmic Impressions: Traces of God in the Laws of Nature* (Philadelphia and London, 2007); quoted from the German edition, *Kosmische Impressionen. Gottes Spuren in den Naturgesetzen* (Vienna, 2004), pp. 48–49.

[7] Ibid., p. 49.

What does faith really teach us? It is at the same time simple yet profound and demanding. If we look more closely, we see that we constantly have to transcend many of our concepts and images in order to penetrate the mystery of creation and to approach it in faith, but likewise with reason.

Let us go back and start again with the first sentence in the Bible. "In the beginning God created the heavens and the earth" (Gen 1:1). "*B^e^reschit bara* . . ." is what it says in Hebrew. This word "*bara*", "creates", is used in the Bible solely for God's activity as Creator. The *Catechism*[8] emphasizes that three things are being said in these first words of Scripture. First, the eternal God has called into being everything that exists outside himself, "heaven and earth". The first sentence of the Bible does not say that God just "blew the whistle at the start", or gave everything a push to start it off. Everything that exists, in whatever way, has been called into being by him. Second, he alone is the Creator. In Hebrew, the word "*bara*" always has God for its subject. He alone can call anything into being. Hence it follows, finally, that everything that exists is dependent upon God, who gives existence to it all.

Three misunderstandings regarding these first three statements have to be cleared out of the way.

The eternal God has called into being everything which exists outside of himself, "heaven and earth".

The first and most common misunderstanding is the way that God is understood as first cause. He is the first cause of all causes, but he does not stand at the beginning of a long chain of causes, so to speak, as if God were the billiard player who sets a ball in motion, and this ball then runs

[8] CCC 290.

and strikes another ball, which rolls and strikes a third—as if God were the first in a long series of causes. Since the Enlightenment, people have been fond of using another image. A clockmaker assembles a clock, and once it has been constructed it then runs until it has to be wound up again. That Richard Dawkins does not want to use that kind of "clockmaker" for our world does not in itself make him an atheist in our opinion.

Steven Weinberg formulates the current assumption of scientific method as follows: "[T]he only way that any sort of science can proceed is to assume that there is no divine intervention and to see how far one can get with this assumption."[9] Scientific method, as understood by Weinberg and many others, involves a conscious renunciation of "divine intervention". This methodical exclusion of divine involvement is sometimes referred to as "methodological atheism". I see it differently: this has nothing to do with atheism but is a straightforward method of natural science. It cannot assume the existence of a "clockmaker" who intervenes. This method is looking for mechanisms and sets of conditions that can explain the way things happen.

We believe in a Creator who is not just one cause among others, intervening from time to time, when things get too difficult or when you reach some limit. God does not intervene in the way that a mother intervenes in her children's squabbles, whenever the game turns into a quarrel. God is sovereign in relation to his creation, of course, and he can, for example, heal a cancerous growth in sovereign fashion through his creative power—we call this a "miracle". We will return to this topic. Here, however, we are concerned with the act of creation as such. It is the sovereign act of

[9] Steven Weinberg, *Dreams of a Final Theory* (New York: Pantheon Books, 1992); p. 247.

the constitution of being as such. "God spoke and it was". Everything that is, exists thanks to this word, this creative act of God. He has created everything, heaven and earth. Everything that exists, on earth and in heaven, visible and invisible—and we also believe that there are invisible creatures, the angels—is a created reality. That is the first and most important assertion.

And there is a question to be interposed here: Is this exclusively a matter of faith, or is everyone able to perceive this with his reason? On this point, the *Catechism* says the following:

> Human intelligence is surely already capable of finding a response to the question of origins. The existence of God the Creator can be known with certainty through his works, by the light of human reason, even if this knowledge is often obscured and disfigured by error. That is why faith comes to confirm and enlighten reason in the correct understanding of this truth.[10]

Fundamentally, our reason is able to recognize that things are created, even if we only receive complete enlightenment about creation through revelation, in faith. Reason can recognize that the world does not exist of itself. Everything is contingent; nothing created itself. (I am leaving the much discussed question of the self-organization of matter to one side, for the moment.) We have made neither the world nor ourselves. We can only, so far as our little capacities allow, change what already exists: for the better, and sometimes, unfortunately, for the worse. But we always have to have something to start from: namely, that we and this world exist. The contingency of creation may seem painful, yet in fact it is not humiliating for man to be dependent

[10] CCC 286. See also Vatican Council I, can. 2, section 1: DS 3026.

upon the Creator; rather, it gives us undreamt of opportunities. On the other hand, it means that the Creator is sustaining everything and that we are being held within his hand.

God alone is the Creator.

Everything we may observe in the world, all making of things, is the moving about and changing of what is already there. The cabinet maker makes a table out of wood; he changes the wood, he shapes it. Using the material already available, he gives it another form. The housewife makes, from a pile of indefinable ingredients, a marvelous meal, forming something already there into something new. Yet it is not something absolutely new that comes into being thereby; rather, it is a re-shaping. It is no different with an artist, a technician, or someone engaged in creative thinking. Even my best ideas are not absolutely new. They always presuppose that other people have previously thought things, and that I have already thought things. The ideas come from the exchange of ideas, and when I have some particular inspiration, then it is always just a re-shaping of what already exists.

God's act of creation occurs without any movement. He does not shape something that already exists. According to most creation myths, the gods create the world by re-shaping what is already there. They are "demiurges", giving form to the chaos, to what is already there, to primal material. Only God, as we meet him in the Bible, is a Creator. Early Christian writers were already discussing in their writings the many ancient creation myths in their cultural environment. Thus, for instance, around 180 A.D. Saint Theophilus of Antioch writes:

> What would be remarkable if God made the world out of pre-existent matter? Even a human artisan, when he obtains material from someone, makes whatever he wishes out of it. But the power of God is revealed by his making whatever he wishes out of the non-existent.[11]

God creates "out of nothing". That does not mean that "nothing", "the non-existent", is something that he makes things out of; it means rather that God's act of creation is a sovereign constituting of things. God spoke, and it was! Everything that is, has been called into existence by him. That is what is wonderful and unique about the biblical belief in creation.

With all their abilities, men can only change what already exists. Yet perhaps something really new does arise sometimes. That is a question we will take up again further on: Is there anything new in the world? What does it mean if, for instance, new species occur in the course of evolution?

Everything that exists is dependent on God.

Belief in creation says that God does not create in time, at some time or other, at a given point on a time-line. Creation is not a temporal act. We only know everything on a time line: yesterday, today, tomorrow, the beginning and the end. God's act of creation is not the first act in a long temporal series, which was done once and then that was the end of it, as if God had done his job and could now sit back and watch.

"In the beginning God created ..." This beginning is always in God's eternity. For us creatures, there is always a temporal beginning. I began sixty years ago. The universe began fourteen billion years ago. Yet God's act of creation

[11] *Ad Autolycum* II, 4.

is not in time—it is he who creates time. He is eternal, and his creating does not occur at given points in time, now here and now there. He calls things into being and he keeps them in being. Creation is happening now, in the "now" of God. In the Letter to the Hebrews, we read, "He upholds the universe by his word of power" (Heb 1:3). That is why we have to say that if God were to let go of creation, even for a single moment, then it would immediately collapse into the nothingness from which he called it up. In order to comprehend this, we have to try to get beyond our temporal and spatial conceptions.

Creation and the Freedom of God

God creates in utter freedom—nothing compels him to do it. He is not acting from necessity, as we do. We are always needing something: needing to eat, or to sleep, in need because we lack something or because we want to realize something, or to realize ourselves. God does not need to realize himself. Creation is not an extension or addition to him, not a piece of himself. We have been freely constituted by him, which means that we have been willed by him.

This has massive consequences for the way we understand the world and ourselves. Because God created us in sovereign freedom, he therefore has given his creatures genuine independence. The creatures are themselves; they really do have their own existence and activity, the autonomy that has been vouchsafed them. The greatest of miracles is the freedom that God gave to man. Before looking at the consequences of this, we have to distinguish it clearly from three other possible solutions, which are all interrelated.

The "emanationist" solution claims that the world is a "spill over" from God, so to speak, a "piece of him" that is of lesser value, "falling-away", a lesser form of God. The "pantheistic" solution sees everything as being in God, and as God. God is in everything, but in such a way that everything is God, even trees and animals. The "monistic" solution says that there is only one substance, one being, and that is God; everything else either does not exist at all, or it is God.

All three "solutions" can be found in esoteric literature even today. They share one basic error in common: they imply that God is no longer God, and that the creatures are no longer independent, but merely a "part of God". These three solutions appear to be pious, and hence they may easily deceive people. They appear to elevate creatures to divine status. In reality, it is the other way round.

Because God created the world and all the creatures in sovereign freedom, without any kind of pressure or obligation—that is, he freely granted them existence and activity—creation has true independence. If the creatures were an "overflow" from God's being, then they would not be independent, and would have no existence of their own. Precisely because we were created by God quite freely, we are able to be truly ourselves.

That has far-reaching consequences. In an "evolutionist" view (which, on an overall worldview, I clearly distinguish from the scientific theory of evolution), people have difficulty in accepting that creatures really can be themselves. Everything is merged into the stream of evolution, nothing seems to have any purchase. Everything is just a snapshot in the river of time. Belief in creation, however, sees in all creatures beings with their own existence, their own form, their activity, and—in the case of men—their freedom.

It follows—and this is quite central—that because God creates in complete freedom, his creatures are independent. They have their own status and their own continuing existence, because they are willed by God. Saint Thomas expresses this as follows: God gives things not only existence, but also their own activity. That is realized to the highest degree in man: we are creatures who have not only received existence, but also mind, will, and freedom.

I know of no other doctrine that so illuminatingly and persuasively combines the dependence of all creatures upon their Creator with the independence of the creatures. This is only possible because God creates in sovereign freedom. He has no reason for creating other than his goodness, which he grants to his creatures to share in. That is what "And God saw that it was good" means.

The Christian belief in the Creator, then, is something quite different from the "deist" notion of a clockmaker, who merely sets something going at the start. Belief in the Creator takes nothing away from the creatures. It gives freedom along with dependency, however paradoxical that may sound. Being dependent upon God is no dishonor, no derogation of rights. Having received everything from him constitutes the dignity of the creature. Belief in the Creator is thus the best guarantee for the protection of his creatures' dignity. If everything is merely the result of chance and necessity, it is hard to see why any dignity should be attributed to creatures, or why they should be given any consideration. And with this last remark, we have introduced the next question: Do creatures have dignity and value, "each after its kind"?

III.

"He Created Everything according to Its Kind" (Genesis 1:11)

What does it mean, in practical terms, that God created "everything after its kind", or "all kinds" (Gen 1:11–12, 21, 24–25)? Was it a separate act of creation each time? For centuries, the belief was widespread that the species were unchangeable, and that each one had been separately created by God. In the nineteenth century, the notion of the "transformation of species" came to be held, according to which the various species had developed gradually and ever higher.

At the end of his "Introduction" to the *Origin of Species*, Darwin sums up the essence of his theory as follows:

> I am fully convinced that species are not immutable; but that those belonging to what are called the same genera are lineal descendants of some other and generally extinct species, in the same manner as the acknowledged varieties of any one species are the descendants of that species. Furthermore, I am convinced that Natural Selection has been the most important, but not the exclusive, means of modification.[1]

Through an honest and intense intellectual struggle, Darwin freed himself from the view, widespread at that time, "that each species was independently created".[2] In a letter to his friend Joseph D. Hooker, in 1844, he wrote of this struggle, "It is like confessing a murder". Thereupon, Darwin entered the nineteenth-century public arena and enjoyed an incredible success. Many people today are very sensitive and as if hurt—even agressive—whenever anyone doubts the theory of evolution. The debate has proven, however,

[1] Charles Darwin, *On the Origin of Species by Means of Natural Selection, or the Preservation of Favoured Races in the Struggle for Life*, 2nd ed., 1859 (London: J.M. Dent & Sons, Ltd., 1971), p. 20.

[2] Ibid.

that there is still a good deal of room for questions, and that we do need to leave that place for questions. People who are questioning and inquiring do a good service, for nothing is worse for science than forbidding questions and inquiries.

A Look at the Biblical Message about Creation

In what follows, I attempt something daring. The creation narrative in the first chapter of Genesis is not a scientific text in the sense modern science would understand. Yet I should like to put questions to the text about how its basic message addresses our thinking, and thus how it is also important for our dialogue with science.

> And God said, "Let the earth put forth vegetation, plants yielding seed, and fruit trees bearing fruit in which is their seed, each according to its kind, upon the earth." And it was so. The earth brought forth vegetation, plants yielding seed according to their own kinds, and trees bearing fruit in which is their seed, each according to its kind. And God saw that it was good. And there was evening and there was morning, a third day. (Gen 1:11–13)

That is the work of the third day of creation. On the fourth day, sun, moon, and stars are created, the "lights in the firmament of the heavens"—even though light had already been created on the first day of creation, as it is so beautifully transposed in the *Creation* oratorio by Joseph Haydn. On the fifth and sixth days, the water and land animals come into being, then finally man. Let us read several passages,

> And God said, "Let the waters bring forth swarms of living creatures, and let birds fly above the earth across the firmament of the heavens." So God created the great sea monsters

> and every living creature that moves, with which the waters swarm, according to their kinds, and every winged bird according to its kind. And God saw that it was good. And God blessed them, saying, "Be fruitful and multiply and fill the waters in the seas, and let birds multiply on the earth." And there was evening and there was morning, a fifth day.
>
> And God said, "Let the earth bring forth living creatures according to their kinds: cattle and creeping things and beasts of the earth according to their kinds." And it was so. And God made the beasts of the earth according to their kinds and the cattle according to their kinds, and everything that creeps upon the ground according to its kind. And God saw that it was good.
>
> Then God said, "Let us make man in our image, after our likeness; and let them have dominion over the fish of the sea, and over the birds of the air, and over the cattle, and over all the earth, and over every creeping thing that creeps upon the earth." So God created man in his own image, in the image of God he created him; male and female he created them. (Gen 1:20–27)

It is not the aim of this text to give us information as to how this world originated. It is not a text on natural science. Saint Augustine already discussed this point, writing very clearly in his treatise against Felix the Manichean:

> We do not read in the Gospel that the Lord said, "I am sending you the Holy Spirit, that he may teach you about the course of the sun and the moon." He wished to make people Christians, not astronomers! The knowledge that men may learn in school, for practical purposes, is enough for them! Christ did indeed say that the Holy Spirit would come, to lead us into all truth, but he is not talking there about the course of the sun or the moon! If, however, you think that the teaching (about these things) is part of the truth that Christ promised through the Holy Spirit, then I ask you, "How many stars are there, then?" I maintain that this kind of thing is not part of

> Christian teaching .... Whilst you maintain that part of that teaching is also how the world was made, and what happens in the world."[3]

Saint Augustine was quite clear about this: We can quite safely and confidently leave the investigation of "How?" questions to natural science. Ought we not, therefore, make a neat, clean division: here on one side, the faith, its documents, the Bible, magisterium, and reflection about this in theology; and there, natural science with its methods, hypotheses, theories, experiments, observations, and results? Of course, they cannot be divided so neatly and clearly. For faith and science both have to do with life. As early as 1959, the great theologian Karl Rahner called on theologians to take note of the fact that they can "not proceed on the assumption that scientific and theological questions and knowledge can have no points of contact."[4] The same thing holds good, however, for scientists. That is why I cannot accept calls from the side of the scientists to keep out of these questions. Certainly, I am not competent as a scientist, but I think it is good for us when we put questions to one another and help each other forward by an exchange of ideas.

The Bible does not offer any theory about the origin of the world and the development of the species. Yet natural science's way of looking at the origin of species is not the only approach to reality. This has to be emphasized, time and again: there are various approaches to reality—philosophical, artistic, religious and scientific. Each one is no less real than another, for they are approaches to the same reality. I want to try to highlight a few "statements about reality" from the

[3] *Contra Felicem Manichaeum* I, 10; PL 42:525.

[4] Karl Rahner, Preface to Paul Overhage, *Um das Erscheinungsbild des ersten Menschen* [Concerning the appearance of the first man], Questiones Disputatae 7, (Freiburg im Breisgau, 1959), pp. 11–30.

first chapter of Genesis. Corresponding to the seven days of creation, I am going to select seven points.

Everything Is Created

Everything that is, is created. That is the first, fundamental statement the Bible makes about reality. The world, nature, does not exist "of itself". Nothing of what exists has created itself. The question as to how far there is something like "self-organization", is much discussed today among scientists.[5] But nothing has any being and activity of itself. The saying of Paul the Apostle, addressed to men, is equally valid for creation as a whole: "What have you that you did not receive?" (1 Cor 4:7) That is the first truth to which the Bible bears witness. And, in principle, it is accessible to reason.

Variety Is What God Wills

The world offers an enormous variety of creatures—the stars, the variety on our own planet, the world of plants and animals. The basic message about this variety is that it is good, it is what God intends, it corresponds to the Creator's will. For anyone brought up on the Bible, this sounds obvious, yet in the history of thought it has by no means always been clear. In the history of human thought, there are two tendencies concerning the question of variety, which come to quite different conclusions. There is one tradition of

[5] I merely refer to one book by Erich Jantsch, an Austrian astrophysicist who taught at U.C. Berkeley, *Die Selbstorganisation des Universums. Vom Urknall zum menschlichen Geist* (Munich, 1979); *The Self-Organizing Universe: Scientific and Human Implications of the Emerging Paradigm of Evolution* (New York, 1980).

thought that says that the variety is the sign of an "accident", an "original accident", so to speak. Originally there was only unity, "the one", which broke up in the accident. As if in a cascade, the one dissolved itself into variety. This tradition of thinking, which may be found above all in Neoplatonism and in Gnosticism, the so-called "Gnosis", is still widespread today. Variety, for this tradition of thought, is a sign of decay, of decadence. The further you get from unity, the more various the world becomes, but also weaker, right to the outermost edge—according to Neoplatonism, as far as material, which is thus reckoned as being negative. In this tradition of thought, it is always a matter of returning from the variety to the unity again. The variety has to be lifted up into the unity once more. Similar tendencies, so far as I can judge, are to be found in the Buddhist view of the world, where the variety of the world, "Maya", is regarded as a deception, as illusion.

There is another way of looking at things, which has become very popular through evolutionism—so popular, that in our sensibilities nowadays, it has become almost a kind of self-evident "truth" that cannot be questioned. The whole variety of living things, according to this view, is not the expression of a reason that sets things in order, of a will or a "plan of creation", but results from the interplay of chance and necessity. Through random variations and their chances of survival in the struggle for existence, there arises a great variety, which develops out into all the corners and crannies of the world, just as far as life can in any way develop itself. It comes about that one random variation finds it has a good chance, a niche so to speak, so as to survive and assert itself. Darwin, in his most famous book, argued against having recourse to individual acts of creation so as to explain the variety of species. He wanted to work out as much of a "natural explanation" for the origin of species as possible.

Even if Darwin had the impression of committing a "murder" here, because he believed he had in some sense to overcome his inherited religious beliefs, this kind of notion is entirely legitimate. The method of natural science looks for natural causes, and it tries to explain situations as completely as possible by natural causes. This methodological limitation is indeed the reason for the success and enormous achievements of this very method. The danger lies, however, in people forgetting the limitations of this method. It shows a narrow segment of reality with great clarity, but we should not regard it as being the whole of reality.

The biblical view shows us a variety that is neither accidental nor random, but rather the expression of the nature and the will of God. Saint Thomas Aquinas asks himself whether the multitude and variety of things in this world spring from God. He discusses the theory of random events, already known in the ancient world and held by the "atomists", according to whom the variety in the world is the result of a random interplay of matter (Democritus). Thomas insists, on the contrary, that the variety is the personal intention of the Creator—God desired a multifarious world:

> For he has brought things forth into being, so as to share his goodness with creatures, and to represent it through creatures. And because it cannot be adequately represented through one creature, he brought forth many and varied creatures, so that what one creature fails to represent of the divine goodness is supplemented by another. For goodness, which in God is simple and has only one form, is variously present in creatures.[6]

No one creature alone can reflect God. It needs the whole plenitude of creatures to represent God's plenitude. The variety of creatures is the multiform expression of the goodness

[6] *Summa Theologiae* I, q. 47, a. 1.

of God. This has one fundamental consequence: as a result of belief in creation, creatures are to be seen in a positive light. At one place in the Book of Wisdom, it says, "Thou hast loathing for none of the things which thou hast made" (Wis 11:24). All creatures have their own value, their own kind of rightness. Every creature, whether it be a star or a stone, a plant or a tree, an animal or a human being, reflects the perfection and the goodness of God in its own particular fashion. They all have their own value and likewise their own effect on the world. Belief in creation lays the foundation for a basically positive view of creation and of all its manifestations. Later, in chapter 5, we shall have to ask ourselves why there is so much that is cruel and negative in this good creation.

Evolutionism as a way of seeing the world (not as a scientific theory) has far greater difficulty with this. For this worldview, there are not really any species, for things have no existence of their own. What we regard as "species" are in fact merely "snapshots" in the great stream of evolution. Everything is just transition and a stage being passed through, and each individual is merely a fluke, which had the luck to survive because it was "more fit" than the others. This is certainly a short-sighted view of the variety of creation. The way men marvel at the variety of nature gives us a hint of something different. Above all, it seems to me, evolutionism as a worldview cannot actually offer any reason why anything has any value in itself, if everything is, so to say, merely a transitory stage in the stream of evolution.

Order in Variety

According to the biblical creation narrative, it is the case that variety is ordered. The question as to what exactly a "kind" or species is, has always been difficult. What does "each according to its kind" mean? Even if it is difficult, in some border areas, to draw clear boundaries between the species, yet the two great spheres of plant life and animal life are quite clearly different. In the Bible, there is a quite clear distinction drawn between plants and trees, on one hand, and the swarm of living things in the water, birds in the sky, and animals of every kind, on the other—right up to man. It may well be a conviction based on experience that there are species: a cat is not an elephant, and a dog is not a mouse. Certainly, a tree is not a bird, and man is not an ape (and vice versa).

Yet what constitutes the various species? Did God create each one individually, the daisy and the ginko tree, the crocodile and the squirrel? Here we come up against the perennial question of human thought, which even evolutionism cannot evade: we can only ever consider single, concrete individuals—this dog, and that spruce tree, this grasshopper, and that man. "Humanity" is not something we can see, nor is "catness" or "spruce-ness".

Behind these considerations lies the perennial dispute about "universals". Is there really such a thing as "humanity", or are these just "nomina nuda", as Umberto Eco says in the final sentence of his famous novel *The Name of the Rose*? Nominalism, which was widespread in the fifteenth century, says that we cannot actually know anything properly. Is there such a thing as "man" as a kind of creature, a species? I have the impression that many scientists do not really like this question because it is too philosophical. It leads us unavoidably into metaphysics. Is there such a thing as a "species"? Are there such things as "beings" at all? The question is not purely academic. We shall meet it again and again, for instance, in dealing with the question of whether man is really "the crown of creation". Here we come up against the question of whether man is essentially different from the animals, or whether he is nothing more than a random variant type of animal. Similarly, the question arises as to the transition from lifeless forms of matter to living things, and again concerning the transition from the sphere of plants to that of animals. In everyday life, we presuppose in each of these cases that there is a difference in the nature of things. Meat, which we eat, comes from animals. Someone may regard eating meat as not very nice, or unhealthy, or reprehensible. Nonetheless, the butcher is still allowed to sell meat and to kill animals for that purpose. He is not allowed to kill humans. A vegetarian gives up eating animals and killing them but not the consumption of plants. A head of lettuce that has been cut off will live no longer. Yet that is quite rightly regarded as interfering less with nature than slaughtering an animal. Plants, animals, humans: these three realms are essentially different, even if in border areas it is not always quite easy to distinguish. We need this way of seeing differences in nature in order to recognize what things are at all.

When we say with the Bible that God created man in his own image and likeness, then we are also saying that he made something essentially new that differs from the plant and animal realms, even though all three are profoundly related to each other. Yet how can we give an account of our faith such that God's creative will is shown to be expressed in this ordering of things? Is Saint Francis of Assisi's "Canticle of the Sun" just a pious exaggeration, or is it addressing some kind of reality? Is it merely a play on words to please pious souls, or is creation, with all its realms, its orders, and its species a "progetto intelligente che è il cosmo", a "rational project" or an "intelligent plan that is the cosmos", as Pope Benedict XVI stated in his General Audience on November 9, 2005?

Darwin was trying to show that the origin of the species is to be understood as a development from a first seed. This fascinating view has become popular, yet it is still full of great questions.[7] The famous philosopher and theoretician of science, Sir Karl Popper, said:

> Neither Darwin, nor any Darwinian, has so far given an actual causal explanation of the adaptive evolution of any single organism or any single organ. All that as been shown . . . is that such explanations might exist (that is to say, they are not logically impossible).[8]

Darwin himself says, at one place in his great book,

> Why then is not every geological formation and every stratum full of such intermediate links? Geology assuredly does not reveal any such finely graduated organic chain; and this, perhaps, is the most obvious and serious objection which can be urged against the theory.[9]

This passage is talking about the famous "missing links", the intermediate forms that are missing. If it is true that everything developed from one first seed, then there ought to be innumerable transitional stages, but noone has yet discovered any of them.

Evolution and Ascent

The creation narrative portrays an "ascent" of species: first the plant world, next the animal world, and then man; an "ascent" from plants to trees, from the "beings swarming in the waters" to the birds and the animals of dry land, and

[7] See *On the Origin of Species*, 400.

[8] Karl Popper, *Objective Knowledge: An Evolutionary Approach* (Oxford: Oxford University Press, 1972), p. 267.

[9] *On the Origin of Species*, pp. 292–93.

finally to man. The theory of evolution, likewise, has a "movement of ascent": from the first single celled creatures, by way of fishes, reptiles, land animals, and apes, right up to man. Why does biological variety increase with time? A friend of mine who was a biologist used to say that it would be much more probable, according to Darwin's theory, if in the end only viruses and bacteria survived since they are much better adapted for the "struggle for existence" than are higher creatures.

Why does man stand at the end of this "ladder of ascent"? Only we can look back down this ladder, on which we are standing at the highest rung, and see that the path that has led to us has a significance. This path has a goal, and yet the theory of evolution finds it hard to cope with the question as to whether evolution is directed toward a goal.

Creation or Nature?

Meditating upon Holy Scripture seems to lead, time and again, to the alternative between accepting the existence of a Creator and ascribing all development to him, or on the other hand attributing everything to purely natural and material causes. Darwin seemed to be standing before this choice: either a Creator or chance. That is what he says himself, at the end of his book:

> Authors of the highest eminence seem to be fully satisfied with the view that each species has been independently created. To my mind it accords better with what we know of the laws impressed on matter by the Creator, that the production and extinction of the past and present inhabitants of the world should have been due to secondary causes.[10]

[10] Ibid., p. 462.

But are these the only alternatives: either a Creator or natural causes? Let us offer just a little observation about the text of the Bible on this point. Genesis says that God addresses the earth, "Let the earth put forth vegetation", and so it "brought forth vegetation" (Gen 1:11–12). God commands the waters, "Let the waters bring forth swarms of living creatures" (1:20), and finally God tells the earth to "bring forth living creatures" (1:24). Does that not mean that God can also work through the earth? All this is pointing toward something that is an essential part of the Christian understanding of creation. The Creator endows the creatures not only with existence, but also with effective activity. He grants them being "without presuppositions", so to speak, in creating them out of nothing. Yet his creatures become fellow creators, through his giving them the laws, the powers, and the capacities to be active. We men can be his fellow creators. Without a doubt, this is what is sublime about the biblical and Christian notion of creation.

What Darwin called "secondary causes" can thus perfectly well be reconciled with belief in creation. The natural causes are an expression of the activity of creation. This can be made clear in one "obvious case", to which we all owe our existence: our parents' cooperation, in procreation, with the Creator who has created us. We believe, and we openly declare, that every person is directly God's creature. Each person owes his "self", his existence as a person, to the Creator, who has called him into existence for his own sake. And yet, the necessary condition for our coming into existence is that our parents have conceived us. Here we can see how Darwin's "secondary causes" are interconnected with the activity of the Creator. There is a profound and mysterious relationship here. And yet, does it not make sense to assume that this happens at all the stages of creation?

The Common Bond of Creatureliness

All creatures are related to one another through the one Creator, by whose hand all creatures have been brought forth. There exists therefore an indissoluble solidarity between creatures. Even man is "only" a creature—he shares that in common with the fly and with water. This basic feeling that we are all united with the whole of creation is strongly characteristic of the pathos of the Darwinian model. Darwin speaks of the "community of descent . . . the hidden bond which naturalists have been unconsciously seeking."[11] He believed he could discover this bond better without having recourse to the activity of the Creator. In the genealogical union of all living beings, he found an inspiring concept with which we wholeheartedly agree. There is something wonderful and exalting about becoming aware of these common roots.

I suspect that modern philosophy, since René Descartes, has radically separated man from the rest of nature, and has set him over against it as a thinking being. With Darwin, man is brought back into being part of nature again. He is a child of the same nature that has produced everything else. However, Darwin takes a step in the wrong direction when, in the process of integrating man with nature, what is particular about man is leveled down.

Creation for the Glory of God

The right direction is indicated by what the creation narrative reveals about the seventh day, the sabbath. Creation has a goal. With man, creation can attain to the knowledge

[11] Ibid., p. 400.

of its Creator. It can recognize him and praise him. With the sabbath, the goal of creation is clearly expressed. The *Catechism of the Catholic Church* says it simply and clearly: "The world was made for the glory of God." [12]

One of the most stimulating opponents of ideological Darwinism, in my opinion, was the great Swiss zoologist Adolf Portmann, who agreed with Darwin on many things with respect to his observations, but he also clearly specified what was deficient. The deficiency that stands out above all is what Joachim Illies in his biography of Portmann calls, "Darwinism's mania for expediency".[13] The world of living things is full of "beauty that has no purpose, perfection that has no practical value, existence of a particular kind offering itself without any value for selection—inexplicable, and thus meaningless (and thereby a scandal) for the well-ordered world of the mechanistic interpretation of reality".[14] This beauty without purpose, the instances of marvelous perfection that are never seen, that have no practical purpose, but are simply manifestations of beauty, "unselfishly" pouring themselves out—we only begin to understand their significance when we look at the goal of creation, which is to praise God.

Perhaps this "purposeless" superabundance of beauty reveals more clearly what criticism of evolutionism as a materialistic view of the world entails. Perhaps music too, which is strictly mathematical like natural science, can help us to look beyond the too restrictive horizons of materialism and thus open us to the melody of the Creator.

[12] CCC 293.

[13] Joachim Illies, *Adolf Portmann: Ein Biologe vor dem Geheimnis des Lebendigen* [Adolf Portmann: A biologist confronting the mystery of living things] (Freiburg im Breisgau, 1981), p. 159.

[14] Ibid., p. 160.

IV.

“He Upholds the Universe by His Word of Power” (Hebrews 1:3)

Does it make sense to pray for good weather? At the end of the sixties I heard a theologian who maintained that it was quite meaningless to pray for good weather, since the weather was entirely determined by causes within this world. Does it make any sense for the children and husband of a mother who is ill with cancer to pray for her healing? Is it, then, an "intervention" by God if she gets well again, or are natural powers at work to heal her? Yet if she is not healed, then what kind of a God is it who ignores the tears of the children and the husband's pleas? Can God not help? Then he is powerless. Does he not want to help? Then he is cruel and merciless.

God Is at Work in the World—A Poetical Approach

Believing in God also means believing that he is active—not just here and there, and not just at the beginning, but constantly. Everything has its origin in him; he upholds everything and gives it its goal. Is this belief simply an arbitrary assumption, a kind of drug with which to anaesthetize ourselves in this difficult world, "the opium of the people", as Karl Marx called religion? Or has this belief firm points of anchorage that show it to be rational, meaningful, beautiful and good? Psalm 104 is indeed beautiful and expresses with great poetry the feelings about nature that many people experience:

> "Bless the Lord, O my soul!
> O Lord my God, thou art very great!
> Thou art clothed with honor and majesty,

Who coverest thyself with light as with a garment,
Who hast stretched out the heavens like a tent. . . .

Thou didst set the earth on its foundations,
So that it should never be shaken. . . .

Thou makest springs gush forth in the valleys;
They flow between the hills,

They give drink to every beast of the field,
The wild asses quench their thirst.

By them the birds of the air have their habitation;
They sing among the branches. . . .

Thou dost cause the grass to grow for the cattle,
And plants for man to cultivate,
That he may bring forth food from the earth,

And wine to gladden the heart of man,
Oil to make his face shine,
And bread to strengthen man's heart. . . .

When thou sendest forth thy Spirit, they are created,
And thou renewest the face of the ground.

May the glory of the Lord endure for ever,
May the Lord rejoice in his works,

Who looks on the earth and it trembles,
Who touches the mountains and they smoke!

I will sing to the Lord as long as I live;
I will sing praise to my God while I have being.

May my meditation be pleasing to him,
For I rejoice in the Lord." (from Ps 104)

Yes, "may my meditation (or poem) please him," sings the Psalmist (Ps 104:34). Is it "merely" poetry if we rejoice in the Creator and his works? Or does this meditation, this song of praise to the Creator nonetheless refer to a reality, to the Creator's activity that is the foundation of everything and upholds everything? To put the question another way, is this approach of poetry less real than the approach of natural science? Let us hear what Sergei Bulgakov, the great Russian theologian and philosopher of religion, has to say. In the following passage, he talks about his "path home" to faith after ten years of wandering in the wilderness of scientistic atheism:

> I was in my twenty-fourth year, but it was ten years since faith had been torn out of my heart, and after crises and doubts, a religious vacuum had taken possession of it. Oh, how dreadful is this sleep of the soul, which may last all one's life! With my mental growth and the acquisition of scientific knowledge, my soul had declined into self-satisfaction, becoming blasé and vulgar .... Suddenly, *this* happened...
>
> Evening was drawing on ... We were driving through the southern steppes, covered in the spicy scent of honeyed grass and hay, glowing golden in the soft light of the sinking sun; in the distance, the first mountains of the Caucasus were already turning blue. I was seeing them for the first time. I looked at the mountains with great interest, I breathed in the air and the light: I was hearkening to the revelation of nature. My soul had long become used to seeing nothing in nature but a *dead wilderness*, covered with a veil of beauty, as if it were wearing a deceptive mask. And suddenly, my soul was filled with joy, and trembled with enthusiasm: *what if there were* ... , what if there were no wilderness, no mask, and no death, but Him, the gentle and loving Father; if that were his veil, his love ... , if the pious sentiments of my childhood, when I lived with Him, when I stood before His face, when I loved Him and trembled at my being incapable of drawing near to Him, when my tears and my youthful ardour, the sweetness of prayer, my childish purity—which I used to joke about, after I had soiled

> it—if all this were true, and the other thing—the vacuum clothed in death—were nothing but blindness and lies? Yet, was this possible? Did I not know, since my years in the seminary, that God did not exist? Could there be any doubt about it? Could I admit to myself having had these thoughts, without feeling ashamed of my cowardice, without a feeling of panic and fright in the face of "science" and its verdict? . . .
>
> And you again, O you Caucasus mountains. I have seen your ice glittering from one sea to the other; your snows reddened by the morning sun, your peaks rising to pierce the heavens; and my soul melted in ecstasy. . . .
>
> The first day of creation dawned before my eyes. Everything was light, all was filled with peace and echoed with joy. There was no life and no death, simply an eternal and immutable present. And an unexpected feeling arose within me, and swelled up high: the feeling of victory over death.[1]

What kind of reality is being hymned here? Does the poetic/religious approach open up a different sphere of reality, which has nothing to do with the one that natural science is interested in? Karl Rahner says in one place, "Theology and natural science are fundamentally unable to enter into contradiction with each other, because they each differ from the other both in the sphere that is their object and in their method."[2] The two do not have to enter into contradiction. Yet the spheres that are their object are not so different that they can avoid contact with one another in practical terms. I am convinced that such contact is necessary and raises no contradictions. Even the poetic/religious approach has to make contact with the scientific approach at some point.

[1] Sergei Bulgakov, *The Unfading Light* (Moscow, 1917), pp. 7–11; cited by M.-J. Le Guillou, *Das Mysterium des Vaters* (Einsiedeln, 1999), pp. 210–11.

[2] Karl Rahner, "Naturwissenschaft und christlicher Glaube" [Natural science and Christian belief], in *Schriften zur Theologie* [*Theological Investigations*], vol. 15 (Zurich, 1983), p. 26.

Yet why all this "anxiety about contact"? If it is true that the Creator is constantly sustaining, upholding, and renewing his work, if all the new things that make their appearance in creation have had (and still have) their origin in his plan of creation and his creative power, then there must be points of contact here with the reality that is the object of the natural sciences. How is this to happen, however, without poaching on one another's preserves—or conversely, without simply refusing to have anything to do with one another?

Creation and Providence

The following considerations are concerned with the *creatio continua*, the continuing creation, which is the same sphere of reality that natural science investigates. In order to avoid a possible misunderstanding, let us affirm from the outset that the reality of continuing creation cannot be measured by empirical methods. Accepting it, however, does not contradict the scientific way of looking at things: it is neither irrational nor impossible to grasp. Believing in creation as a current event, now taking place, not only makes sense but is also the presupposition for science having a meaningful basis. Nevertheless, such a claim still requires a great deal of explanation, and the reasons for it need to be made clear.

First of all, another approach will throw the question of continuing creation into sharper relief. The first approach was connected to prayer, and the content of the second was the beauty of creation, which becomes a way of approaching the Creator. A third approach is what I should like to call the "existential" one. This plays a central role in the preaching of Jesus. It concerns belief in divine providence; not in an abstract general providence, but a quite particular one. Jesus teaches his disciples that they can trust absolutely in this practical, providential care, which enters into the smallest details, on the part of someone Jesus calls "the heavenly Father". In the Sermon on the Mount, we read:

> Therefore I tell you, do not be anxious about your life, what you shall eat or what you shall drink, nor about your body, what you shall put on. Is not life more than food, and the body more than clothing? Look at the birds of the air: they neither sow nor reap nor gather into barns, and yet your heavenly Father feeds them. Are you not of more value than they? And which of you by being anxious can add one cubit to his span of life? And why are you anxious about clothing? Consider the lilies of the field, how they grow; they neither toil nor spin; yet I tell you, even Solomon in all his glory was not arrayed like one of these. But if God so clothes the grass of the field, which today is alive and tomorrow is thrown into the oven, will he not much more clothe you, O men of little faith? Therefore do not be anxious. (Mt 6:25–31a)

In another place, Jesus says, still more clearly:

> Are not two sparrows sold for a penny? And not one of them will fall to the ground without your Father's will. But even the hairs of your head are all numbered. Fear not, therefore, you are of more value than many sparrows. (Mt 10:29–31)

If Christians believe not just in a general providence, but also include in this a particular providence, for even "the hairs of your head are all numbered", what does this imply for atoms, molecules, and all of matter? We cannot avoid this question if the preaching of Jesus and scientific research are not to part company. Unless this belief and the scientific way of looking at things are to stand side-by-side without any contact with each other, then this question confronts us with considerable challenges in conceptual terms. Belief in creation and the approach of natural science can best complement each other, without each trying to speak and decide on behalf of the other. But such a complementarity pressuposes some intensive thinking.

Evolution as a Matter of Belief?

Let us refer again to a contemporary phenomenon. The actor Thomas Kretschmann, who played Pope John Paul II in an American television production, is supposed to have said, "I have nothing to do with the Church. I do not believe in God, I believe in evolution, that's more logical to me."[3] Is one's attitude to evolution a question of faith? The Christmas issue of *Der Spiegel*[4] carried the title "Gott gegen Darwin: Glaubenskrieg um die Evolution" [God versus Darwin: a religious war over evolution].

How has this strange "sacralization" of a scientific theory come about? How is it that this theory is the only one, so far as I know, that has become an "-ism"? There is no "Einsteinism" corresponding to Einstein's theory of relativity, nor is there any "Newtonism" or "Heisenbergism". Why is there a "Darwinism"? American philosopher and historian of science Stanley L. Jaki has said that freeing Darwin's theory of evolution, and its further development in neo-darwinism, "from what is not science there", so that it does not become ideology, but remains science, is a most important task.[5]

Anyone who makes a "battle of beliefs" out of the question of evolution is certainly not serving science. The fact that questions concerning evolution have been made into

[3] *Der Standard*, January 5/6, 2006, p. 5.

[4] December 24, 2005.

[5] Stanley L. Jaki, "Non-Darwinian Darwinism", in R. Pascual, ed., *L'Evoluzione: Crocevia di scienza, filosofia e teologia* [Evolution: crossroads of science, philosophy and theology] (Rome, 2005), pp. 41–52, at p. 41.

"weapons of war" to use against belief in creation, has little to do with science, just as Marxism's "dialectical materialism", with its "scientific" atheism, has very little indeed to do with genuine science.

Anyone who seeks to get beyond slogans and prejudices will have to make a great intellectual effort. Yet the trouble is worth taking.

Continuing Providence

We will approach the belief in continuing creation in three conceptual steps. It cannot be proved, but we can certainly show that this belief does not contradict reason. Before attempting these three steps, we should point out once more what *creatio continua* is not. The German theologian Ulrich Lüke, who deals with this subject in detail, asks whether continual creation, in comparison with the creation at the beginning (*creatio ex nihilo*), "is the awareness of a task of improvement that the great constructor, the Creator of things out of nothing, assigns himself. Is *creatio continua* the 'maintenance agreement', that had to be concluded, as it were, at the time of delivery, the *creatio originalis*, the beginning of creation, with respect to the product's quality?"[6] The notion that "current creation" means that God has to "re-adjust" and repair his creation, is common. If the present activity of the Creator is understood as a kind of "process of improvement", then it is understandable that people only want to have it involved in places where there are gaps in our knowledge, as a kind of "stopgap" for areas scientific knowledge has not yet reached.

[6] Ulrich Lüke, "Creatio continua", in *Theologie und Glaube*, no. 86 (1996): 281–95, at p. 283.

I should like to indicate a different path here. The first step is a step backward, so to speak, to what happens in the everyday activity of natural science. It is a philosophical consideration of "contingency", which is nonetheless also of existential importance for our lives.

A great deal that was previously incomprehensible in natural processes, because we did not know how to explain it, can be explained today through scientific research and has thereby become comprehensible. God the Creator does not appear in these explanations, however, but "merely" matrices of material causation. The more that is explained, the less there remains that is inexplicable. Is the "room" for God becoming steadily "smaller"? It is no wonder that *Der Spiegel* closes the article we just cited with the words, "It's becoming cramped for the creator."

Yet belief in the Creator does not begin at the point where we do not yet know something, but precisely where we do know very well. The proper approach is to look at what we already know today. That, thank God, is a great deal. We are not looking where there is still something unexplained to see if there is still room for God, but looking at what we know and asking, "What is this based on?"

One thing is certain: every observable being at one time did not exist. The sun came into existence, as did the moon, the earth, life in all its forms, right up to man, and myself as well. What once did not exist will also decay and disappear again. What once came into existence has no existence of itself, and its state of existence is unstable. It is then meaningful and necessary to ask, "What keeps everything in being?"

Nothing that has a material existence exists "necessarily". It could equally well not exist: the sun might not have come into existence. The same is true of me. I am here because I have come into existence. Philosophy calls this "contingency", the absence of necessity in our existence. What keeps us in existence, then—why are we here? Why do we not lapse into nothingness? Psalm 104 replies, "When thou hidest thy face, they (the creatures) are troubled; when thou takest away their breath, they die and return to their dust" (v. 29). This keeping-in-existence is what philosophy and theology call "continuing creation". Everything that is, God keeps in being. Without this support, it would not exist. The power that maintains everything in being cannot be yet another material force. Such a force would have to be supported by something in turn, and that again by something else, and so on to infinity. That is why the Jesuit philosopher Rainer Koltermann says, "The power maintaining things, ultimately, can only be something that is not itself kept in being by something else." It can be no force that has come into being and is ultimately finite; no energy that we can measure. It must be an absolute power, beyond time, infinite. "These characteristics are essential for God."[7]

We call this power *creatio continua*, the continuing activity of the Creator. It is this activity that "holds the world together in its inmost being". If God were to let go of creation, then it would fall back into that from which it came, into nothingness.

A further thought follows from these considerations. Not only is the existence of all things upheld by this original source, the power of the Creator, but also the activity of all things is sustained by the "original activity" of the Creator. For the activity of all things is likewise "contingent", not necessary—it might equally well be different. The final cause

[7] Rainer Koltermann, *Grundzüge der modernen Naturphilosophie: Ein kritischer Gesamtentwurf* [An outline of a modern philosophy nature: a critical sketch] (Frankfurt, 1994), p. 134.

for the creatures' capacity to be active cannot be another activity within the world, a finite, created energy. God effects "all things in every one" (1 Cor 12:6), says Paul, but not in such a way that he is one cause alongside others. Rather God is the cause that maintains all activity whatsoever and makes it possible. That is how we may understand that powerful saying from the Letter to the Hebrews: "He upholds the universe by his word of power"—everything that exists, and everything that is active (Heb 1:3).

New Things in Creation

But what is it like when we are concerned with a creative act that is more than just a "maintaining in continual being"? What is God's creative activity like when it is concerned with the appearance of something genuinely new: for instance, the appearance of life—and especially, of man? Does God cause the "leaps" from non-living to living things or from animals to man? Have we come back again to the individual "acts of creation", which Darwin thought his theory of natural selection rendered superfluous as an explanation?

Let us dare to take a second step. There is no doubt that our world is a world of becoming, in which the unfolding of the cosmos, and evolution, have made human life possible on our planet. Along this path of becoming, however, we find the "appearance" of genuinely new things. Can this "greater thing" have arisen from the "less"? Can lower things bring forth, of their own power, higher and more complex things? Nothing in our experience suggests that something lower can give rise to something higher, simply of itself, without some directive and organizing activity and, still less, do so quite by chance.

Are there, then, "individual acts of creation" after all? Yet how could we establish their existence? Here we need to refer to a perfectly simple distinction that people prefer to overlook. This is the distinction between a precondition and a cause. In order for life to come into being on our planet, a whole series of preconditions were needed, without which there would be no life. Yet these preconditions were—and are—only the framework of conditions for life to come into being. They do not constitute the creative cause of life. They all play a part in life's coming to be; yet the new element in the development of the world, which we call life, cannot be derived from them. For it to come about, it truly needs the creative act of God, the "divine spark", in order to come into being.

This "divine spark", this "let there be . . . and it was" (Gen 1), natural science will never encounter on its own plane of operation. It is trying to apprehend ever more precisely, in ever more complex ways, the conditions necessary for new life to be kindled in the process of development. Because research into the conditions requisite for life is making such enormous progress, many people believe they have unlocked the secret of life itself. As elements that genuinely make this possible, these requisite conditions are in that sense contributory causes, but they do not create life. I can try to clarify this through two examples from human life, in which what I mean can perhaps be grasped.

The preparation of catechetical lectures is associated with concentrated reading and thinking. Ideas are brought together and clarified in discussion. Then the process of writing them down begins, and finally comes the lecture. All that requires many preconditions without which it is not possible. One's brain has to be more or less working, and there has to be time for preparing the catechesis. The organs of sense have to serve their parts, and equally the pen; paper will be used, as will the microphone in the cathedral. All of these are requisites or conditions, they help the thing to happen, but they do not make up the catechetical lectures. The new thing that comes into being here (and even that is not absolutely new) needs these prerequisites, but is not produced by them. Neither the pen nor the microphone, not even the brain of the person speaking, have made these lectures.

Much the same is true of evolution. The great "leaps" by which the stages of evolution ascended each had their necessary preconditions, which cannot however be the things that created these new realities. They are genuine contributory causes, but not the actual creative cause. The theologian Cardinal Leo Scheffczyk wrote, "Thus evolution becomes in a sense comprehensible as creation that does not exclude or eliminate creative co-operation, but brings it wholly into play: for in this way of conceiving it, the act of becoming something new presupposes the presence and activity of creaturely reality, with all its proper forces, its dynamics, and its causality. This, then, is a communal act in which God and the creature both participate."[8]

The second example makes it even clearer. It concerns the highest instance of correspondence between creaturely conditions and the divine act of creation: the coming into existence of a new human being. If it is true that each human being is unique, then this cannot refer merely to being genetically unique, in the way that no human being is entirely identical with another. It is much more concerned with the uniqueness of the person, which is most clearly expressed in the unconditional dignity of the human

[8] Leo Scheffczyk, "Gottes fortdauernde Schöpfung" [God's continuing creation], in idem, *Schwerpunkte des Glaubens: Gesammelte Schriften zur Theologie* [Focal points for faith: collected writings on theology] (Einsiedeln, 1977), pp. 177–205; esp. p. 200.

person. Every man possesses this dignity, as a human being, independent of his descent, sex, achievements, or health. We say that this dignity is attributable to man as a creature "in God's image and likeness" (Gen 1:26).

There is no doubt that the parents are real contributory causes of the existence of a new child. Yet they do not "produce" this new person. The really new thing that enters the world with the new human child comes into being, to use Scheffczyk's words again, through "a communal act, in which God and the creature both participate",[9] not however "on equal terms", but so that the parents are wholly a cause, through their contribution of everything that is proper to them, at their own level, and so that God is entirely the cause of the new person, by making him—which is only possible through the divine act of creation: a new person, immortal in his soul, unique in his calling before God and to God.

In this view of divine causality, God's activity is not that of a *deus ex machina*, a "stopgap", who has to be brought in for whatever is "not yet explicable". It is not a matter of "intervening, case by case" from outside, but of the transcendental creative activity of God, who alone makes it possible for this world to "hold together", and for it to climb higher, step by step, in accordance with his plan, for genuinely new elements to appear in it, right up to man.

Anyone who wants to replace the Creator in the realization of this plan by a complete autonomy of evolution, inevitably either attributes a mythical creative power to evolution itself, or renounces any attempt whatever at rational comprehension by explaining everything as the blind interplay of arbitrary chances—which, as I wrote in the *New York Times*, would be "to make reason redundant".

[9] Ibid.

The Incredible Wealth of Variety

A third step may shed a little more light on the meaning of "continuing creation". Let us look at some comparisons that help us to interpret the incredible variety of creation. Where does this absolutely limitless and effortless variegation in the forms of life as well as inanimate creation come from? Are they all merely "expedient variants"? Many are, certainly, yet there is more than mere usefulness involved. The more we research into nature, the more strongly we have the impression that there is at work a prodigal delight in multifariousness, in beauty—and equally, in what is bizarre, frightening and sinister—which does not appear to follow any rationale of purpose. I always have a nagging suspicion that the Creator also simply enjoyed playing with this variety. Thus, I dare to express the thought: Could it be that this enormous variety exists as a result of God's inexhaustible creativity?

I was helped here by thinking of Wolfgang Amadeus Mozart. All his works are "contingent", for they might equally well not have been written. They were mostly written for particular purposes, as commissions, ordered by other people; yet many were also written simply out of a brilliant creative impulse—even when these were commissioned pieces. Purpose and beauty are not dissociated here. A work of art may have a purpose, yet it is more than its purpose. Artists do not create new things "out of nothing". They use existing models. Composers follow existing harmonies and musical rules. They take up themes and melodies from other people, and develop them further; and yet the result is unique. Mozart simply developed music further, and yet he produced unique works. We marvel at Mozart, we love and revere him. It would not occur to anyone that these pieces of music had put themselves together.

In comparison with that, the doggedly ideological and materialistic evolutionism seems to me truly sad and unimaginative. Would it not be a good thing to look at the theory of evolution in the light of the creative power of someone like Mozart, for once? Would we not be coming closer to the Creator, as he plays his inexhaustible melodies in his creation?

V.

"You Guide Everything . . ."— God's Guidance and the World's Suffering

The poet and writer Reinhold Schneider spent the winter of 1957–1958 in Vienna. It was his last winter. Only fifty-five years old, he kept a diary over these last months. Suffering great pain and subject to profound melancholy before he died on Easter 1958, Schneider spoke again and again about the incomprehensible cruelties in nature, about the "process of eating and being eaten"[1] and also about the senselessly horrific world of men, filled with suffering and wars and groundless wickedness.

Critical Questions Put to Belief

Had the sick and depressed poet lost his faith, which had supported so many people during the Nazi era? Had he returned to the tragic view of the world that had influenced him before his conversion to the Catholic Faith? His reflections, helplessness, and uncertainty, bordering on despair, in the face of the cruelties of this world, call into question the belief in a good Creator, his meaningful plan, and his loving providence. We may quote three entries from these diaries. On the occasion of a visit to the museum of natural history, Reinhold Schneider remarks:

> Let someone just walk through the museum of natural history—God is just as close as he is distant. It is impossible to deny him in the face of this immense world of forms, this frightening plenitude of inventions; to deny him, in the face of the absurd architecture of a dinosaur—a cathedral of meaninglessness, of a will to live that is incapable of living; in the face of

[1] Reinhold Schneider, *Winter im Wien* [Winter in Vienna] (Freiburg im Breisgau, 1958), p. 184.

> Japanese crabs like evil spirits, a long-legged loving couple out of hell; of the octopus, an eightfold combination of head and feet—which, if I remember rightly, people confronted with a giant lobster, in the Hamburg aquarium, for the edification of visitors. The course of the encounter was surprizing: the octopus wrapped himself around his opponent's claws, broke them up, and sucked the life out of its shell. And the starfish breaks up mussels, thrusts its stomach tube in, and drinks them right out like an egg. About the sharks, which attack the walruses—from the side; about the defenselessness of seals and dolphins, there is nothing to say; nor anything about the fights of giant jellyfish with whales; about the frog, whose life is sucked out by a leech entwined around him, while he is standing upright like a human being . . .[2]

Why are there parasites, and their unimaginably cruel activities, in a creation which is good?—Or is it just our imagination that makes us shudder? Is "nature" simply like that, perhaps, pitiless and unsympathetic "cycles of hell", as Schneider puts it?[3] Let us listen to him again:

> We have to pray, even if we cannot. I can very well pray for others, for priests, researchers, statesmen, nations, for creatures, for the earth; for sick people first of all, as is obvious, and for the dead; that is the silent affirmation of a mysterious connection. I have a profound need to do this; this is what keeps me going, what calls me to church in the morning; for myself I cannot pray. And the face of the Father has become quite overshadowed. It is the dreadful mask of the one who smashes things up, the one who is treading the winepress. I simply cannot say "Father" . . . . Let us read a chapter about parasites (in the books by Natzmer, by K. von Frisch—who does nonetheless truly observe lice, bugs, and fleas with

[2] Ibid., pp. 129–30.
[3] Ibid., p. 171.

> the eyes of love, an almost unique instance—or of L. von Bertalanffy). Let us recall the everyday tale, already many times retold, of the parasites who live in the intestines of certain birds, and whose eggs find their way into snails by way of the droppings; within the snails, the seeds grow out into tubelets that make their way up into the horns; within the horns, as these swell up, they develop an attractive color pattern and equally attractive movements; these tempt birds to peck off the horns; and thus the parasites come back to their own place again. And the snails are constantly growing new horns, and these are constantly being torn off; the snail is only a provider for the destructive parasites who are destroying it and the birds. Unless it is host to a myriad of destructive parasites, and unless it allows them to make use of it, no higher organism could exist; without them, then, intellect could not express itself. And what of love and beauty, then? It requires extraordinary power never to damage them.[4]

We could find as many other examples as we liked. Who has not heard of the "praying mantis", the female of which eats the male alive, during sexual coupling? Where is "intelligent design" there? Where is the kind and loving Creator who can say of his creation that it is good?

Finally, what ought we to think about the never-ending concatenation of human suffering? Reinhold Schneider wrote down what he had read in the newspaper in a single day about meaningless and random cases of children's suffering:

> Time expresses itself most clearly in absurdities; I cannot help setting it all down together: in Holland, a four-year-old girl was treated with radioactive material; the point of the needle broke off and remained unnoticed in her; the child became ill, and the contamination emanating from her has driven the gallant Haanschoten family from their little house in Putten; the garden is also infected, as is the path to the children's school

[4] Ibid., pp. 119–20.

> that runs past it. In Vienna, an eight-year-old died of operational shock while having a tooth removed by the dentist; in Oakland, California, a court has decided in favor of the parents of two children who suffered incurable paralysis and curvature of the spine after inoculation with the Salk vaccine; four children died in the neighbourhood of Bari, after being injected, on the instructions of the health service, with a serum against diphtheria that had not previously been used; fifteen children are in the hospital. At the University Hospital in Munich, a nurse most unfortunately injected a young girl with benzine, instead of a narcotic sedative; the patient died. This can be read on the same day, just leafing through the newspapers.[5]

Such examples could be multiplied at will. In reaction to my article, I received many letters in which the following question was expressed often: How can a rational plan of creation be found in a world full of absurd chance happenings? A professor of medical computer science, Wolfgang Schreiner, wrote to me last autumn:

> Previously, the theme of evolution was not, for me, an especially explosive subject . . . In the last three years, however, I have been engaged in detailed bio-computer studies. This work necessarily involves going fairly deeply into genome research. Nowadays, we can look at the entire genome on any computer. If previously I had thought that "Creation is well-ordered, and all disorder is only a deviation from this well-ordered plan . . ." (whether or not someone is to blame for it), then very soon I necessarily had quite the opposite impression: of absolutely unplanned steps. Creation seems more like an accumulation of unplanned steps, and we see only those "products" that have survived (and evince a certain degree of functional use). Hence this is described (by many interpreters) as having been "shaped according to a plan". This is just like someone describing his win in a game of chance as the result of a planned-out pro-

[5] Ibid., pp. 126–27.

cedure . . . . Would that convince us? One can of course always interject here, "It does appear to be unplanned, but only because we do not understand the plan underlying it . . ." Even this argument is one I wish I could hold to be true.

But have a look at the genome for yourself, Cardinal! See how it goes "all over the place". It is much more like a town that has been improved many times, where new things are built up on part of the ruins of the old. Copies and imitation copies of things have been introduced in suitable and unsuitable places, partly turned in the right direction, and partly the wrong way around. No technician would ever plan such a muddle. Indeed, it gives the impression of being just "the opposite of planning".

For me, as a Christian, this knowledge was profoundly disturbing; I had hitherto held the exact opposite to be true . . . Ultimately, in my view it points in this direction . . . : God made use of evolution to create all of this. Granted this, however, the real problem is now on the table: How can God, the merciful one, allow all those frightful attempts and false trails with thousands of deaths? And all of these things are supposed to be the means used by his planned creativity? It contradicts the common picture of God that we have received (passed on by the Church). . . .

It is certainly high time for us to look carefully at the world that God has made in order to find out how he really intended it to be. I think no one yet knows.

In Search of an Answer

But are there answers to these difficult and urgent questions? The question concerning the origin of pain and its relation to the goodness of God should by no means be given too hasty an answer. Saint Augustine struggled with this question. He writes, "I asked about the origins of evil, but could find no way out."[6] Only after seeking for a long time and trying many circuitous paths and false trails did he find an answer; he found the One who alone is victorious over evil, sin and death.[7] And thus it is clear that "only the Christian faith as a whole . . . can be the answer to this question."[8]

Centuries of Christian experience plainly show that the more deeply we are taken into the infinite mystery of God, the more alive is our fellowship with Christ, the Redeemer, the more intimately we are acquainted with the Holy Spirit, and thus the more clearly the question concerning evil in the world will emerge. On the other hand, the more our sense of God's presence fades, the more clouded our understanding of evil will become. It will become a meaningless scandal, in the face of which one can only take refuge either in despair or in denial. Ivan Karamazov, in the novel by Dostoyevsky, *The Brothers Karamazov*, takes refuge in revolt. In view of the suffering of children, Ivan draws the conclusion that the world and history are absurd. He wants to return his "membership card". For him, a God who lets people suffer pointlessly cannot exist. Dostoyevsky himself is struggling with the question as to whether there can be any answer at all to the atheistic arguments based on the dreadful suffering of innocent creatures. Nor will it be convincing for this weighty question simply to point out that we can only succeed in showing what the answer may be through a change in our point of view, our perspective. Dostoyevsky sketches an answer to Ivan's protest-atheism in the figure of the staretz, Zosima. Arguments ultimately cannot convince anyone that the ills of the world are not just a pointless absurdity. It has always

[6] *Confessions* VII, 7, 11.
[7] CCC 385.
[8] Ibid., 309.

been people who have lived out the convincing answer. Mother Teresa, for example, was just such a living answer to the challenge of pain and evil in the world.

Nonetheless, we have to develop arguments as well. Reason wants to understand why evil exists in the world. The answer from the great Christian tradition of thought is profound and carefully thought through, and there is an urgent need for us to know it better. The most important points are summarized in the *Catechism of the Catholic Church*.[9]

The Best of Worlds?

Let us begin with one of the most frequent objections to the Creator and his plan, his guidance and his providence: "Why did God not create a world so perfect that no evil could exist in it?"[10]

One misunderstanding is widespread. It proceeds from the assumption that when God created this world, he can have created it only in complete perfection. Every failing, then, that we observe appears to weigh against the idea of a rational Creator and his intelligent plan. The "muddle" in the genetic code is just such an example. People then like to say that no rational engineer would construct a machine in that way. Another example is the human eye. As a "naïve believer in creation", I should say that it is an inconceivable miracle that arouses my astonishment at the Creator's work. Experts in evolution however, retort that no optician would construct the objective, the lens, and the reflector according to the way they are now arranged in the human eye. Before I address the misunderstanding

[9] Ibid., 309–11.
[10] Ibid., 310.

underlying this approach, just a "quick" argument in reply: it may be that the human eye could be better constructed. It is thanks to this "construction", however, that we are able to become opticians, engineers, and other things—indeed, that we are able to experience the marvel of vision, so far as no defect hinders us in this. And in spite of all our fantastic technical expertise, no one is capable of constructing a human eye that is alive and working.

But let us turn to the heart of the question: Does God have to make a perfect and faultless world if he is the Creator? Is the alternative either a perfect creation or a world that is purely the result of chance? Must God as Creator make a world in which everything has already been brought to completion, in which everything, right from the beginning, is perfectly formed? Would he be obliged to do so?

If it were the case that creation means having a beginning, followed by a process of development, which finally reaches a goal, then this would after all mean that with what he "created in the beginning", the Creator launched the world on a path that it is still following now, and the end of which has not yet been attained. In such a world, things would be constantly coming into being. Yet together with that, there is always equally a passing away that is involved. For no material thing that comes into being and develops can remain static. It will always likewise pass away again. In a "world of becoming", there is inevitably also decay, destruction, and death. The *Catechism* therefore concludes, "With physical good there exists also *physical evil* as long as creation has not reached perfection."[11]

Let us for the moment stay with "physical evil". In the discussion about creation and evolution, this plays a greater role than the question about moral evil, which only arises

[11] Ibid.

in the sphere of created freedom—that is, in the case of men and of angels (cf. chapter 6). Can the "God saw that it was good" of Genesis be reconciled with the "eat and be eaten" in the entire world of living things? Let us try to approach this question in three stages.

To start with, we need a kind of "fundamental metaphysical insight". The fundamental principle runs thus: "Everything that is, is good, because it exists". It is good to exist. The natural sciences always involve characteristics: size, quantity, quality, origin, place, time. We are asking, however, about the vehicle of all characteristics and defining points. Before all characteristics, it holds true that everything that exists has being. And that is good. And following this, there is an important clarification from belief: God gave everything its being. This does not mean, however, that everything that exists is therefore already "the best possible". God could equally well create a better world.[12] He could do so, if that were his will. Yet would he not be obliged to do it if he could do so? We are easily offended at this point: Why, if he could have done so, did he not create a better, more just, and more lovely world? Is he perhaps not powerful enough to create a better, more carefully planned world? Did he "lose his grip on it"? Does he not have adequate access to things or not enough power? Or does everything go according to chance, without any meaning or purpose? Is it not easier to understand without a Creator?

Everything That Is, Is Good—But Limited

When Michelangelo had finished his Moses, he is supposed to have thrown his hammer at it, saying, "Well then, speak!"

[12] Cf. *Summa Theologiae* I, q. 25 a. 6, ad 1; CCC 310.

No work of art, however perfect it may be, can express or exhaust everything that the artist has in his heart or in mind. It will always be a limited expression—limited by the material framework that always sets limits to all ideas.

No creature—this is the second step—can entirely express its Creator. Even if the world were far better, far more perfect, it would never come anywhere near God's glory. It always remains only a reflection of the greatness of the Creator. This is also connected with the fact that all creatures are in a process of becoming, that they have a beginning, a development, and an end.

The never ending debate, as to whether there is something like a "design" in creation, thus goes round in circles, perhaps because nowadays, whenever people talk about "design" and a "designer", they automatically think of a "divine engineer", a kind of omniscient technician, who—because he must be perfect—can, equally, only produce perfect machines. Here, in my view, lies the most profound cause of many misunderstandings—even on the part of the "intelligent design" school in the U.S.A. God is no clockmaker; he is not a constructor of machines, but a Creator of natures. The world is not a mechanical clock, not some vast machine, nor even a mega-computer, but rather, as Jacques Maritain said, "une république des natures", "a republic of natures." [13]

In order to talk meaningfully about the Creator having a "design", we have to retrieve the concept of "nature", an understanding of which we have largely lost today, and which has been replaced by a technical and mechanistic understanding of living things. To say that God creates a

[13] Jacques Maritain, "Reflections on Necessity and Contingency" in *The Range of Reason* (New York, 1952); quoted from the French edition, "Réflexions sur la nécessité et la contingence", in idem, *Raisons et raisons* (Paris, 1947), p. 62.

"republic of natures" means that in his creation there is above all what is called *phyein* in Greek, growth and becoming, with all its hesitation, its trials, its failures and its breakthroughs, its instances of cooperation and of conflict, its inconceivable prodigality and its unexpected by-products—both successful and unsuccessful. All natures have their own unmistakable form of activity and influence, with which they are endowed by the Creator, and which lets them grow out of themselves and be active. They reach their goal or attain their purpose not by a force applied from outside them, but by one from within them, working outward. Saint Thomas Aquinas says that a "nature" is an "inner principle" on the basis of which everything does those things, and has those effects, which correspond to its nature. He attributes this inner principle to the *ars divina*, the art of the Creator, who endows the creatures with their self-development and the way they shape themselves.[14]

Let us return to the question of the prodigality of nature. "A fig tree, in its lifetime, which often lasts over a hundred years, produces tons of seeds, one of which will finally be successful, when a gale has got rid of the old tree and has made room for new growth. Until then, the tree will have made food available for birds, or wild boar, and insects," the geneticist previously quoted wrote to me, and he believes that this "principle of over-production", and the destruction associated with it, "cannot be a glorious chapter in the story of the activity of any God". For an engineer, this contradicts the principle of being technically purposeful. But does it contradict the vitality and creativity of life? This limitless prodigality is connected with the need to assure survival, but it is also a sign of life, which with all its

[14] Cf. Thomas Aquinas, *In Physicorum* II, lec. 14, no. 8.

imperfection and its transitory nature is nonetheless a reflection of how inexhaustibly alive the Creator is.

Can Evil Even be Good?

The decisive question—and with this we reach the third step—still needs to be tackled: *unde malum*?—Where does evil come from? We may perceive that creation is good, even though it is imperfect. Yet why is there so much meaningless destruction and cruelty in it as well?

Natural catastrophes like earthquakes, volcanic eruptions, or tidal waves, remind us, time and again, of how violent and destructive "nature" can be. As powerful as the force of the movement of the continental plates was in 2004, a force that produced the tsunami in the Indian Ocean, the planet earth, our home in the universe, was hardly affected. The rate at which the earth turns slowed down by two-millionths of a second. Dreadful as were the consequences of the tsunami unleashed by the submarine earthquake, the occurrence of earthquakes is merely the reverse side of something that is an essential condition for life on our planet. Without plate tectonics, the mobility of the plates that form the earth's crust, there would be no life upon earth. The experts say that this is one of the preconditions for the earth's being able to maintain a stable average temperature over billions of years—something without which life could not have evolved.[15] The earth is the only planet in the solar system that possesses this flexible geological structure. It is also the only place in which higher life forms have been able to develop.

[15] Peter Ward and Donald Brownlee, *Rare Earth: Why Complex Life Is Uncommon in the Universe* (New York: Springer-Verlag, 2000), pp. 191–220.

We are confronted by a paradox: the thing that causes earthquakes, which time and again bring about the deaths of many people, is at the same time the precondition for the existence on this earth of ourselves, and of all complex life forms. What does this tell us? Marco Bersanelli, an Italian scientist, offers the following reflection:

> A little wave in the ocean, or an almost imperceptible breath on our planet's skin, is enough to devastate our lives. Such phenomena show how fragile and delicate the world is—the world we take for granted, day by day. Normality in the universe is by no means a peaceful sea that is swarming with life; it is, rather, the limitless wilderness of silent worlds, or the unleashing of irresistible forces. The explosion of a supernova near to us would bring about our total and instantaneous annihilation, and yet the same explosion, in the distant past, provided us and every organism with elements that are essential to us, like hydrocarbons and acids. Life on earth exists in an extraordinarily protected niche, marvelously set apart, in which the products of the entire history of the cosmos are exploited.[16]

After the tsunami, time and time again it was said, "How can a good Creator allow such things?" Unfortunately, I have never heard it said, "How can we thank the Creator for giving us the fantastic beaches of Phuket?" They came about through the same geological forces as did the tsunami. We live on a marvelous planet, but everything on it is equally under threat, and our life here has been entrusted to us "subject to recall".

What is true, by and large, is equally applicable in detail. Wolfgang Schreiner, the professor of medical computer science whom I quoted above, gave me an insight into his

[16] Marco Bersanelli, "Nature as God's Creature: Good but Imperfect", *Traces*, no. 2 (2005), http://www.traces-cl.com/feb05/natureasg.html (accessed September 16, 2006).

fascinating work. The unimaginably complex processes in the genome, which determine all the forms of life, are enough to perplex any reflective researcher. Why is there this inherent susceptibility to errors? Why are there deformities, as a result of harmful mutations? Professor Schreiner thinks that we may talk about a "successful design", since it is successful for the survivors. Yet if we look at the mistakes, he says, it is neither a painstaking, "careful design" nor is it a sympathetic, "compassionate design". He concludes that "one would expect everything from an 'intelligent design'" to be successful, careful, and compassionate.

Pointers toward an Answer

I want to try to indicate two directions in which we may find answers. Evil is great, it is dreadful, and it cannot be explained away. Yet good is nonetheless always greater and more powerful—of this we may be convinced with absolute certainty. Evil, in all its forms, is always a lack of good, a defect, which may be great and frightful, yet is ultimately never greater than the good of which it is a distorted or deprived form. Even the smallest genetic defects may produce frightful malformations. Nonetheless, these are still exceptions. The astonishing thing is that such an enormously complex setup as that of the genome works at all, and for the most part does so extraordinarily well and accurately.

A handicapped child, even if it is only one among a thousand healthy ones, has its own life to live, as a unique being, alongside that of its parents and its brothers and sisters. Why does God "allow this"? Let us beware of "glib" answers. Where the question "Why?" is concerned, only the response of solidarity can convince anyone. I, too, might have been this handicapped child. It has the same human existence and the same dignity as I do. It is a living challenge to me: Be the kind of person to me that you would wish for if you were in my place. How much love has come into the world by this painful means!

There remains, finally, Reinhold Schneider's question, the burdensome question of why evil strikes in such a pointless fashion. There is no "design" to be recognized here; far more, the destruction of any meaning and plan. The Bible is aware that "the whole creation has been groaning in labor together until now" (Rom 8:22). It is not only "subjected to futility" (Rom 8:20), but also characterized by evil. It is as if the "power of evil" had gained the upper hand in creation. Yet there is also a promise that has been made to it: "Creation itself will be set free from its bondage to decay and obtain the glorious liberty of the children of God" (Rom 8:21). This way that "creation waits with eager longing" is directed toward "the revealing of the sons of God" (Rom 8:19). The next chapter deals with man as the "crown of creation" and also with the radical critique of this idea. The subsequent chapter is devoted to Christ, the Redeemer, in whom the sufferings of creation come to an end, and in whom the new creation has its beginning and its goal.

VI.

"... Little less than God" (Psalms 8:6) Man as the Crown of Creation?

Vatican II says, "According to the almost unanimous opinion of believers and unbelievers alike, all things on earth should be related to man as their center and crown."[1] Can this position still be maintained forty years later? Is everything on this earth really aligned in relation to man and supposed to be related to him? Should that be the conception believers hold? Is it also shared by non-believers? The unanimity about man being "the crown of creation" that the Council presupposes here does not seem to exist at all. For this would mean, at the same time, that evolution had reached its goal with man—and, thereby, that it corresponded to some plan. Man's coming to be would then be the result of a purposeful process and not a random one.

For many people today, to consider man as the crown of creation sounds like an arrogant exaggeration of one's own importance. We often read that where the Christian faith once (it is said) raised man up and exalted him above all other living beings, science has now knocked him off his pedestal again.

The manner of speaking about "the three great offenses to humanity" offered by science has become an accustomed scheme of things. The well known naturalist and researcher in behavioral patterns Antal Festetics describes this idea, which derives from Sigmund Freud, as follows:

> The first offense came from Krakow, from Copernicus (the earth is not the center of the universe), the second from London, from Darwin (we are descended from animals), and the third offense came out of Vienna, from Sigmund Freud (with the analysis of our mind or "psyche"). We were most deeply hurt by Darwin's "blasphemy" about being related to primates,

[1] Vatican II, *Gaudium et Spes*, 12.

> and we found it both painful and infuriating, for it is apes in particular who look so much like us and who also "ape" the way we behave.[2]

So as to confirm the way man's dignity has been "offended" by the progress of science, we offer yet another example: a few months ago, scientists succeeded in unlocking the genome sequence of chimpanzees. It is more than 98% identical with that of human beings. "The crown of creation" is tottering and faces strong competition. Is it not better to agree with the English evolutionary biologist Olivia Judson:

> Many people would like to think that we men are the product of a special creation, separate from the other forms of life. I do not think this, I am happy that it is not so. I am proud to know that I am a part of this tumult of nature, to know that the same forces which produced me also produced the bees, the giant fern and the microbes.[3]

Any number of examples could be found. In three ways, man's position as "crown of creation" has been called into question: the earth has been removed from the center of things, into an orbit around the sun, somewhere on the edge of a galaxy of over a hundred billion stars, which again is on the edge of over a hundred billion galaxies in the universe. Man is derived from the realm of animals. In itself this is a problem neither for faith nor for reason, as we shall see. What is offensive here is that he is supposed to reflect in some sense an unbroken line of development from nature, so that between animals and men there is supposedly no discontinuity of nature, no metaphysical distinction. Man, as a being with a mind, is seen as something not radically

[2] *Die Presse*, January 30, 2006, p. 30.

[3] Olivia Judson, "Our Place in Nature's Riotous Order", *International Herald Tribune*, January 3, 2006.

new in the great world of nature. Finally, the third offense is that the mind of man has tumbled down from its intellectual eminence and is revealed as a mask for unconscious drives. It is not the mind but the libido that determines his actions. After being "dethroned" in these three ways, the "crown of creation" is said to be in a certain sense groveling in the dust. If the crown remains in the dust, then science has permanently dethroned man. Is he a king or a knave? What is man? In Psalm 8, we sing:

> When I look at thy heavens, the work of thy fingers,
> the moon and the stars which thou hast established;
> what is man that thou art mindful of him,
> and the son of man that thou dost care for him?
> (Ps 8:3–4.)

Is man a part of nature or the "crown of creation"? Or, is he both? Is he descended from the animal realm or a unique creation of God? Or, is he both? Modern science has pushed him aside to the edge of the universe as a miniscule speck on a tiny planet. Or is he nonetheless the inmost goal of the whole gigantic process of development in our world? Or is he both? Has he been humbled because he is now aware of how he is lost in this vast universe? Or has he been exalted because the one point in the universe that he is, is also the tiny point at which the universe becomes aware of itself and can reflect about itself? Or, is man both? The psalmist continues his hymn of praise:

> Yet thou hast made him little less than God,
> and dost crown him with glory and honor.
> Thou hast given him dominion over the works of thy hands;
> thou hast put all things under his feet,
> all sheep and oxen,

and also the beasts of the field,
the birds of the air, and the fish of the sea,
whatever passes along the paths of the sea.
O LORD, our God,
how majestic is thy name in all the earth! (Ps 8:5–9.)

The World Was Created for the Sake of Man

What the Bible says about man has amply shaped the Christian and Jewish tradition. In the "Letter to Diognetus" (early second century) we read the following:

> For God loved mankind *for whose sake he made the world*, to whom he subjected all things which are in the earth, to whom he gave reason, to whom he gave mind, whom alone he enjoined that they should look upward to him, whom he made in his own image, to whom he sent his only begotten Son, to whom he promised the kingdom in heaven—and he will give it to them who loved him.[4]

This passage reflects an "anthropocentric" view of the world and a "theocentric" view of man. Man is the center and the climax of creation. Everything has been created for him and for his sake. His observable superiority of body and mind (speech, reason, and upright stance) argues for this, as does the special favor he has received from God (as created in his image, as the goal of God's incarnation, and as called to eternal bliss).

Christianity shares this conviction with Judaism. The Talmud contains the lovely parable of the world's having

[4] *Epistle to Diognetus*, 10:2, in *The Apostolic Fathers*, vol. 2, trans. K. Lake, Loeb Classical Library (Cambridge, Mass./London, 1926), p. 371.

been made by God like the nuptial chamber that a father prepares for his son. When he had everything prepared, he led his son into the nuptial chamber.[5]

As beautiful as the picture is, does this glorification of man stand up to questioning? On the threshold of the modern age, a young genius, Giovanni Pico della Mirandola, eulogized the outstanding dignity and greatness of man entirely along the lines of the Renaissance, understood in a Christian sense. He presents God speaking to man and reminding him of his unique status:

> O Adam, we have given you neither a particular habitation, nor any particular face, nor any special gift, in order that you may make for yourself, as it pleases you, whatever dwelling, or appearance, or gift, you choose. All the other beings are determined according to the laws of their nature, which we have prescribed; thereby they are kept within their bounds. You, on the other hand, are not limited by any boundaries; you are rather to determine your own nature, according to your own free will, in the hands of which which I have placed your fate. I have set you in the midst of the world, that you may comfortably look around you here at everything there is in the world. We have created you neither as a heavenly nor as an earthly being, neither as a mortal nor as an immortal, in order that you, Master, yourself, may be granted the honor and the duty of shaping your own nature, in whatever kind of life you prefer. You are just as free to distort yourself into subhuman forms as you are free, by your own choice, to be reborn into higher and divine forms.[6]

[5] *Sanhedrin* 108a, cited by H.L. Strack and P. Billerbeck, *Kommentar zum Neuen Testament aus Talmud und Midrasch*, 9th ed., vol. 3 (Munich, 1994), p. 248.

[6] Giovanni Pico della Mirandola, *Oration on the Dignity of Man*, cited in Christoph Schönborn, *Existenz im Übergang* [Existence in transition] (Einsiedeln, 1987), p. 36.

This already anticipates something of what modern anthropology has especially elaborated: the open and indeterminate nature of man is both a weakness and a strength. His particular characteristic, however, is self-determination, that marvelous and unique possession of freedom. It carries with it the risk of being misused and the chance "to be reborn into higher and divine forms". Along with the whole of Christian tradition, the young Pico sees the unique end of man as lying in "divinization", in becoming like God. This sets man apart from all other creatures into a unique position.

Since the discovery that the earth is no longer the center of things, that the sun does not revolve around it, but vice versa, since the "Copernican revolution", the belief in man's central place had also been seriously shaken. As the dimensions of the universe were opened up by the inquiry and exploration of man, so the Christian belief that the earth held a privileged place, as the home of man—as the central point chosen by God, to which he sent his Son as Savior—became ever less credible. Why should this little planet, and man upon it, play any special role?

The famous trial of Galileo in 1633 came to the conclusion that the heliocentric picture of the world could not be reconciled with the Bible. That was hardly a glorious chapter in the history of the Church. Galileo's legal case, which has been enveloped and transfigured by so many legends, presents for many people, to this day, a symbol of the Church's opposition to science, as if the Church's magisterium wanted to dominate science and prescribe what it was and was not allowed to discover. Faith seems to demand that the earth, and man who lives on it, hold a special place; yet science does not seem to be able to hold this view.

Who is man? Is he "a gypsy on the edge of the universe"? This is what the Nobel prize-winner for physics, Jacques Monod, called man in his famous book, *Chance and*

Necessity. Almost at the same time as Monod was writing his book, the Second Vatican Council was maintaining yet again, with all possible solemnity, the exalted position of man—"man, who is the only creature on earth which God willed for itself".[7] Does man hold a privileged place?

Who is man? Is man a "someone", or a "something"? Is he clothed in ineffable dignity—not because anyone has granted him this, but because he has always possessed it as man, because he is man? Or is man a "thing" who can only properly feel himself to be part of a greater whole? All the great questions relating to human dignity and human rights ultimately revolve around this question. The way we should deal with human dignity and human rights depends on how we answer this question. One thing should be said at the start: the answer to this decisive question cannot be found by opposing faith and knowledge, religion and science, but only in a shared effort of thought, research, and also belief.

Man—A Part of Nature

Throughout the history of ideas, time and again there have been "swings of the pendulum". Some people want to take man right back into being part of nature and dispute that he has any special place at all. Others exalt him too much and isolate him from the rest of nature so that he ends up being strictly opposed to "subhuman" nature. In modern times, it was above all René Descartes, and the way he influenced modern thought that one-sidedly separated man from the rest of nature as "a thinking thing" (*res cogitans*), as he said, and understood everything else in a mechanical way

[7] *Gaudium et Spes* 24:3.

as an "extended thing" (*res extensa*), as a quantitative world. On the other side stand the attempts to let man sink back into nature as a whole. The Christian view, in contrast to these extreme positions, unites both aspects in a balanced fashion. Let us look at this a little further.

Man is a part of nature. But in what sense? A glance at the beginnings of the discussion may illuminate this point. When Judaism and Christianity stepped onto the stage of the ancient world, some pagan philosophers energetically opposed the idea that God had created the world for the sake of man. Celsus, the second-century philosopher, for instance, stated that "The world came into being as much for the sake of animals as for that of man."[8] He says that man is wrong to be proud of his special position. "We men feed ourselves with great difficulty, at a cost in human suffering, whereas everything grows for the animals without their sowing or ploughing."[9] We are almost reminded of Jesus' words when he says, "Look at the birds of the air: they neither sow nor reap nor gather into barns, and yet your heavenly Father feeds them" (Mt 6:26), and "Consider the lilies of the field . . . even Solomon in all his glory was not arrayed like one of these" (Mt 6:28). Was Jesus not pulling man down a little from his overestimation of himself here? Did not Jesus refer, time and again, to the way that God concerns himself with *all* creatures, "even the sparrows"? Celsus argues along quite similar lines when he asks, "Does food grow for men alone—and not, rather, for all living things?"[10] The notion that we are lords of creation simply because animals are subject to our will, he rebuts with the impressive argument that not only do men eat

[8] Cf. Origen, *Contra Celsum* IV, 74.
[9] Ibid., IV, 76.
[10] Ibid., 75.

animals, but animals eat men. The fact that we plan and build marvelous cities is no reason for us to think ourselves superior, says Celsus, "for bees and ants make societies and buildings that are just as wonderful". Finally, "All things were created not for the sake of man, any more than for the lion, the eagle, or the dolphin", since—thus his central argument runs—it is a matter of the *whole.* God created the whole of the world, he says, and God concerns himself with the whole. Each thing within this whole has its particular destiny, its own place, man no more than the ape and the rat, says Celsus.[11]

The view this ancient philosopher opposes to the Judaeo-Christian one is of great current interest: man is a part of the whole—this has been a central argument from ancient times up to today. Immersed in the stream of life, man is no different from other creatures. No mind, no powers, and no special calling distinguish him from them. He should be content with this, it is said, and finally abandon his claim to be "something better".

This view, which reduces man to just a part of the whole, has something fascinating about it, and more and more people enthusiastically—even fanatically—came to agree with it, and still do. A number of twentieth-century totalitarian ideologies, which only have room for the state, the party, the race, or the class, and regard the individual merely as a member of the whole—as a part, but not as an individual agent—have made human dignity and human rights correspondingly subordinate. Ideological evolutionism, which I continue to distinguish clearly from the scientific theory of evolution, is certainly related to Celsus' view, with the single difference that in those days "the whole" was seen as static, whereas it is now seen as dynamic. Everything is one

[11] Cf. Ibid., 99.

single great process, the stream of evolution. Thus, the microbiologist Reinhard W. Kaplan, at the end of his book *Der Ursprung des Lebens* [*The Origin of Life*], draws the following conclusion about the way the world should be seen:

> Thus, today, we no longer see life as something incomprehensible, but as a comprehensible stage of the self-development of matter, and thus as being embedded in the gigantic evolution of the cosmos as a whole.[12]

This statement is not false, but it is certainly one-sided, at least with regard to man. An essential aspect of the phenomenon of man is not mentioned here. Of course, everything on earth—matter, living things, and man—is "embedded" as a whole in the gigantic event of the coming into being of the cosmos. Whether, and in what sense, we are right in calling the process of coming into being "evolution", is another question. What is certain, however, is that we owe our bodily existence to the fact of the world's coming into being, beginning with the elements that originated in the process of the universe coming into existence, and right up to the conditions that made life possible on our "gentle planet".

Immersed in the Stream of Things Coming into Existence

To be "immersed" in the stream of things coming into existence is entirely compatible with the biblical view of man. It is one of the marvelous aspects of our earthly life that as men we are related to all other creatures. We share with

[12] Reinhard W. Kaplan, *Der Ursprung des Lebens* [The Origin of life] (Stuttgart, 1972), p. 252.

them the same laws of matter, the same building blocks of life. We have no other sphere of life than that of all other living beings. We are all together in this "Noah's ark" of a planet.

How profoundly our bodily existence is interwoven in the history of the universe is demonstrated by Arnold Benz, professor of astrophysics at the ETH in Zurich (Swiss Federal Institute of Technology). Matter, the elements, of which our body is made up, originated in the tremendous nuclear fusion processes within the stars:

> The carbon and the acids in our bodies come from the helium-burning area of an old star. Shortly before, and during, a supernova explosion, two silicon nuclei fused to make the iron in our blood. The calcium in our teeth was formed from acid and silicon during a supernova. Fluoride, with which we brush our teeth every day, was produced through a rare neutrino-interchange with neon, and the iodine in our thyroid glands originated through the capture of neutrons in the collapse before a supernova. We are directly related to the development of the stars, and are ourselves a part of the history of the cosmos.[13]

The University of Milan astrophysicist Marco Bersanelli adds: "we are in the literal sense 'children of the stars'".[14] To acknowledge this is in no way humiliating. Being part of the universe is nothing to be ashamed of. In the ancient world, people liked to describe man as a microcosm. The whole of the cosmos is present in him, and he in it. It is fascinating to investigate the connections and correspondences that unite man with the biggest and smallest things—with the infinitely tiny world of the atom and the immeasurably great world of the galaxies.

[13] Arnold Benz, *Die Zukunft des Universums. Zufall, Chaos, Gott?* [The future of the universe: chance, chaos, God?] (Munich, 2001), p. 35.

[14] Unpublished; from the manuscript of a lecture delivered in 2004.

It is likewise no humiliation when it becomes clear that man's entry on the stage of the earth entails a long story. The long road of "hominization" is the subject of intensive research, and the more we know about it the less certain our reconstructions of a genealogical tree have become. Is there a common line of descent? Are there several "origins"? And above all, from what chronological point can we talk about "man"? Can there be a gradual transition from animal to man? If man, *homo sapiens*, developed from "hominids", man-like species, then how did hominids become men?

The anthropologists specify certain anatomical and cultural characteristics by which man's special place can be recognized: the size of the brain, upright stance in movement, the use of fire, the formation of traditions, the manufacture and use of tools, and finally language.[15] How did all this come into existence? What made man into man? Is it because of the genes? Yet if the chimpanzee possesses an almost identical genetic code to that of man, then where is the difference?

The Small Distinction

Does a distinction need to be made then? Many people today do not want to do so. Like the ancient philosopher Celsus, they point to the striking similarities between man and animals that occasionally seem to present animals as superior to man.

A short anecdote will show why such a distinction—despite the close relationship—cannot be denied; indeed, if

[15] Cf. Rainer Koltermann, *Grundzüge der modernen Naturphilosophie: Ein kritischer Gesamtentwurf* [An outline of a modern philosophy nature: a critical sketch] (Frankfurt, 1994), pp. 212–65.

we look honestly, it is obvious. A colleague, a fellow Dominican, used to "entertain" us in the conversation at table every day by talking about his intention to write a philosophical book in which he intended to demonstrate that man was no different from the animals. After he had told us about this time and again, one day it got too lurid for one of the brethren, and he asked him, "Father, is this an autobiographical book?"

Our laughter—and his embarrassed silence—provided a clear answer. There is a distinction, a difference of nature, between animals and man. We do not know at what point in the development that led to man this difference appeared. We do know, with evidence clear to reason, that it exists. What does this difference consist in? In our consciousness? Animals, too, have a kind of self-awareness. In forming relationships? Animals also have a kind of relationship: between each other, and (often quite movingly) with humans. In being persons? Certainly, but what makes someone a person? I especially like the approach the German philosopher Hans-Eduard Hengstenberg has worked out. In his view, the specific quality of human beings is their "capacity for objectivity", that is to say, the ability to look beyond one's immediate vital interests and needs and to be aware of oneself, other people, and other things as they are. I do not merely have feelings; I can also observe my feelings, and I can approach them "objectively" and "work on them". I am not wholly "immersed" in my world, for I can take a look at it and change it, I can compare it and make criticisms of it. I can reflect upon it and upon myself. That, however, cannot derive from organic matter, from material with life in it. That cannot "consider itself", so to speak, and stand over against itself.

To a great extent, chimpanzees and men share the same genome. Yet no chimpanzee would ever take an interest in its genome, still less be able to decipher it. His world is limited to bananas, to reproduction, to his environment and his needs. Man is able to investigate his own genome and the chimpanzee's genome as well. He can take an interest in his relationship to the chimpanzee and study it. He even has the freedom to deny being any different from the chimpanzee. Yet he can only do this because he is endowed with a mind. Only a human being could have the idea of writing books to deny that he is any different from other animals. Even for that, you need mind, reason, and will.

Evolutionism (as an ideology) presupposes this freedom. It is only thanks to the mental capacities of man that one can work out theories deriving those capacities from matter. This is what evolutionary epistemology does when it tries to derive the human capacities for perception and knowledge purely from evolutionary advantages in adaptation and survival. This is what evolutionary ethics does when it tries to explain ethical behavior purely on the basis of what is useful in evolutionary terms. It has been demonstrated often enough that none of these attempts can succeed. In order to think, one needs a brain; but a brain does not produce thinking, any more than a piano produces Mozart's piano concertos. They could not be heard without a piano, yet the piano is only a necessary instrument. In the same way, mind cannot be derived from matter, even though our mental processes require material conditions for their framework.

Herein lies the difference between a materialistic view and one that finds a place for the mind. This is not, primarily, a dividing line between faith and science, but between an irrational and a rational view of things. Materialism cannot be maintained as a way of thinking, for it is inherently contradictory. One can, as a matter of method, of scientific approach, blank out the questions of man's spirit and of reason and look only for material causes and connections. Yet this methodological limitation amounts to a decision of the mind. It is only possible for free agents.

The Choice between Reason and Irrationality

One beautiful example will make clear this refutation of materialism—which is actually the classical one. We may find it in the writings of the Jewish philosopher Hans Jonas. When he was writing his great *Prinzip Verantwortung* [*Imperative of Responsibility*], it became clear to him that it makes no sense to talk about ethics and responsibility if there were no such thing as mind, soul, reason or free will. Genes are not responsible for anything. They are not prosecuted in law when they produce cancer cells. Not even animals are held responsible for anything. Only people bear responsibility and are called to account for what they do.

Any and every business activity is a straightforward refutation of materialism. For in business I have to be responsible for things, unlike ants and bees, which work but are not held responsible for any mistakes. Nonetheless, even intelligent people fall into the error of a materialistic interpretation of man. Here, then, is the example that Hans Jonas recounts in order to refute materialism:

> In about 1845, a group of like-minded young physiologists gathered in Berlin, all students of the famous Johannes Müller, who wanted to transform physiology into an "exact" science; they met at weekly reunions in the house of the physicist, Gustav Magnus. Two of them, Ernst Brücker and Emil du Bois-Reymond, took a formal "oath", "to bring people to recognize the truth that there are no other forces at work in any organism than the common physical and chemical ones". There was soon a third member of the covenant, the young Helmholtz,

> whom they got to know at Magnus's in 1845. All three of them achieved great fame and held true to their youthful aims—with brilliant success in scientific terms. What escaped them was the fact that by making their promise they had already broken its terms .... In the act of taking the oath, they were entrusting with power over the behavior of their brains something entirely non-physical, namely their relationship with the truth—and that was something, the existence of which, in a general way, they were denying in the content of the oath ... Promising something, and believing in one's power to keep the promise—and in the alternative of being able to break it, which is similarly comprehended therein: that makes room for a force, within reality as a whole, which is different from the reciprocal workings of the forces "inherent in matter" which can work on inorganic bodies.[16]

The three scholars were right to take into account in their research only "physical and chemical forces". They were mistaken, however, when they assumed that they had thereby said everything there was to say about man. Their "promise" shows that there is a dimension of the mind, of the soul, of reason and freedom, which cannot itself be a product of the material conditions influencing the workings of the mind.

Yet if the mind of man cannot derive from the material conditions bearing upon it, where does it come from? It is reasonable to assume that there is a mental or spiritual principle within man, which the philosophical traditions usually call the "soul". The soul is precisely what makes man a man. Its existence cannot, of course, be "demonstrated", yet without it, without some principle of mind or spirit that rises above material existence, there would be no science, which is after all a "mental business".

[16] Hans Jonas, "Macht oder Ohnmacht der Subjektivität?" ["On the Power or Impotence of Subjectivity", in *Philosophical Dimensions of the Neuro-Medical Sciences* (Dordrecht/Boston, 1976)] (Frankfurt, 1981), pp. 13–14.

From Socrates onward, philosophers have come to the conclusion that the soul is immortal. Many people have assumed, in consequence, that it must be eternal. The Church, however, teaches that "souls are directly created by God". This assertion comes from Pius XII in his 1950 encyclical *Humani Generis*.[17] The Pope emphasizes here that the notion that the human body has its origin in material that already exists and is living does not stand in contradiction to the faith. The human soul, on the other hand, cannot be a product of evolution. Nor is it "produced" by the parents. It is directly created by God.[18] This doctrine of the Church represents the concrete application of the biblical teaching about the special creation of man, alone of all living things, "after the image and likeness of God" (cf. Gen 1:26). Indeed, man was taken from the earth, according to the second account of the creation, and shaped out of this by God; but he became a living being, a man, only when God breathed into him "the breath of life" (Gen 2:7). He is associated with all other living things through his earthly origin, but only on account of the soul that God "breathed into him" is he a man. This accords him his distinctive dignity as well as his unique responsibility, which exalts him above all other living things and at the same time makes him their shepherd.

"The Fly Was Created before You"

Is man the "crown of creation"? Do the "three offenses" still hold good? It is true that our earth is just a speck of dust in the universe. Yet it is becoming steadily more clear

[17] DenzH 3896.
[18] Cf. CCC 366.

how inconceivably privileged this planet is, and to what degree life on this planet, our home, is an instance of incredible improbability. The earth is not the center of the universe, but we do live on a quite admirable "privileged planet."[19] And we men are—so it seems—the only beings on this planet who realize this and who are able to know more and more astonishing things.

It is true that we are a part of nature, and we take our place in the great process of coming into existence in this world. Yet we know about this, and we can conduct research into our place within this process, we can reflect upon it and draw conclusions from it—responsible or damaging conclusions—with a unique freedom.

It is true that we are also guided by instinct, our behavior determined by basic drives, and yet we can also, through research, make our ideas quite clear about this. We further have the responsibility to rise above our instincts and to come to terms with them in responsible fashion.

When we look more closely, the "crown" of creation has not been dethroned. Our enormously increased knowledge should make us more humble, more thankful, and more conscious of our responsibilities. Two rabbinic sayings may conclude our reflections. Jewish wisdom is often attractive and always has a humorous note. The humor brings us up short, whenever we take ourselves too seriously.

> Why was man not created until the sixth day? In order to be able to respond to him, if he should ever show himself to be too arrogant, "You have no reason to be arrogant—the fly was created before you!"[20]

[19] This is the title of the book by G. Gonzales and J. W. Richard, *Privileged Planet* (Washington, 2004).

[20] *Tosefta Sanhedrin* 7: 4–5, cited by E. Urbach, *The Sages* (Jerusalem, 1975), p. 218.

A second saying runs as follows: "Man weighs as much in the balance as the whole work of creation."[21] From these two there follows a third rabbinic saying, which Father Georg Sporschill, the friend of the street children, likes to quote: "Anyone who saves someone's life is saving the whole world."

[21] *Avot* of Rabbi Nathan, cited by Urbach, p. 214.

VII.

“All Things Were Created for Him” (Colossians 1:16)—Christ—The Goal of Creation

Michelangelo Buonarroti depicted the creation of man on the ceiling of the Sistine Chapel. This picture is world famous. The creative hand of God is almost touching the hand of Adam, his creature. The Creator is portrayed as a powerful person with a long, thick beard who is supposed to be God the Father. This image of God the Father as an old man is found in many pictures of the Trinity.

What we find in the entry hall to Saint Mark's cathedral in Venice is quite different. One of the mosaics from the beginning of the thirteenth century is devoted to the work of creation. Following exactly the account in the first three chapters of Genesis, the creation of heaven and earth is portrayed with its climax in the creation of man (as man and woman), then the fall, and finally the expulsion from paradise. In these mosaics, God the Father is nowhere to be seen. The Creator is Jesus Christ, who can be recognized unmistakably by his figure and his cruciform nimbus. Christ is the Creator of the world. It is similar in the cycle of mosaics at Monreale, in Sicily. The ancient iconographical tradition quite clearly always portrayed Christ as Creator. Not until the Renaissance (with one or two isolated exceptions) was a break made with this tradition, and God the Father as Creator was depicted instead.

Christ—Creator of the World

The cycle of pictures in San Marco faithfully translates into the language of images what the New Testament testifies to many times: "In the beginning was the Word, and the Word was with God, and the Word was God.... *All things were made through him*, and without him was not anything made that was made" (Jn 1:1–3). "The world was made through him" (Jn 1:10). Going beyond this view of faith yet again, there is the powerful paradox, the great mystery, that constitutes the heart of Christian belief: "And the Word became flesh and dwelt among us, full of grace and truth; we have beheld his glory, glory as of the only Son from the Father" (Jn 1:14).

The mystery of the Incarnation held a central place in the thought, teaching, and life of the great Pope John Paul II. His first encyclical, *Redemptor Hominis*, begins with a glance at the Incarnation: "The Redeemer of man, Jesus Christ, is the center of the universe and of history."[1] And then the Holy Father writes:

> This act of redemption marked the high point of the history of man within God's loving plan. God entered the history of humanity and, as a man, became an actor in that history, one of the thousands of millions of human beings but at the same time Unique! Through the Incarnation God gave human life the dimension that he intended man to have from his first beginning; he has granted that dimension definitively—in the way that is peculiar to him alone, in keeping with his eternal love and mercy, with the full freedom of God—and he has granted it also with the bounty that enables us, in considering the original sin and the whole history of the sins of humanity, and in considering the errors of the human intellect, will and heart, to repeat with amazement the words of the Sacred Liturgy: "O happy fault . . . which gained us so great a Redeemer!"[2]

There is probably no passage from the documents of Vatican II that the Pope quoted so often as the following one: "The truth is that only in the mystery of the incarnate Word does the mystery of man take on light."[3] God, world, and

[1] *Redemptor Hominis*, March 4, 1979, no. 1.

[2] Ibid.

[3] *Gaudium et Spes* 22:1.

man—everything, through the mystery of the God-man Jesus Christ, appears in a new light.

What does the Incarnation of God in Jesus Christ say to us about man? If Jesus Christ is the Son of God become man, as the Christian faith teaches, then the unique place of man in the universe as a whole is yet further reinforced and exalted. That is what the Council says: "He [Christ] ... is Himself the perfect man", who as Creator has now become the brother of all men, "For by His incarnation the Son of God has united Himself in some fashion with every man."[4] If man is "the crown of creation", then Christ is "the crown of man".

Does the mystery of Christ the Creator and Redeemer have power also to "illuminate reality", a power that opens up new horizons for the intellectual world of natural science? Or, on the contrary, do scientific knowledge and scientific theories oblige us to "push off the throne" not only man, but above all Christ? What place can an incarnate God hold in the overall evolutionary process? Hoimar von Ditfurth, a writer who specialized in popularizing science and was highly successful in the eighties, held the view that the evolutionary view of the world no longer permitted an "absolute" significance to be ascribed to Christ. This is his argument:

> The absolute value that has hitherto been ascribed to the event at Bethlehem in Christian understanding stands in contradiction to identifying the man who personifies this event with man, in the form of *homo sapiens*. There is general agreement that man in his present-day form is—even from the biological point of view—an imperfect, "unfinished" being. To speak in terms of the history of evolution, he has not yet completed the transitional stage from animal to man, has not yet fully realized himself as a true man.

[4] Ibid., 22:2.

> Is the possibility of identifying with such a being really removed, by historical relativization, for all future time? We cannot clear the difficulty away out of the world by leaving out of consideration, for instance, the existence of our evolutionary descendants, which still lies in the future, and is thus not yet a concrete fact. For "absolute" does in fact also mean, in particular, being independent of all future developments. It means the unalterable significance, for all time, of a concrete historical person, who is is simultaneously supposed to be understood as being *homo sapiens*.
>
> I can see no way of dealing with this contradiction other than to admit that the person of Jesus Christ can also be historically relativized in a fundamental sense. Why, in fact, should this not be possible, without affecting the substance of what alone matters? Working out what formulations might do justice to this problem will have to be left to the theologians. I can only indicate the existence of a difficulty here.[5]

There can certainly not be any way of bringing together science and Christian belief at the price of giving up the central mystery of faith. But is there some way of looking at both the evolutionary view of the world and Christian faith at the same time? We have to mention here Father Pierre Teilhard de Chardin, whose controversial work has for quite a while been intellectually and spiritually fascinating. He died in New York on Easter day, 1955.

"In Him All Things Hold Together"

Let us start with what we find in the Bible. In several passages in the New Testament, Christ is praised as the Creator, as the one *through whom* God has made everything. Besides

[5] Hoimar von Ditfurth, *Wir sind nicht nur von dieser Welt* [We do not belong to this world alone] (Munich, 1984), pp. 21ff.

the prologue to John's Gospel, we should mention the hymn in the first chapter of the Letter to the Colossians. In the following passage, Christ is seen as having in some sense cosmic dimensions:

> [Give] thanks to the Father, who has qualified us to share in the inheritance of the saints in light. He has delivered us from the dominion of darkness and transferred us to the kingdom of his beloved Son, *in whom* we have redemption, the forgiveness of sins. He is the image of the invisible God, the first-born of all creation; for *in him* all things were created, in heaven and on earth, visible and invisible, whether thrones or dominions or principalities or authorities—all things were created *through him* and *for him*. He is before all things, and *in him* all things hold together.
>
> He is the head of the body, the Church; he is the beginning, the first-born from the dead, that in everything he might be pre-eminent. For *in him* all the fullness of God was pleased to dwell, and *through him* to reconcile to himself [literally, *toward him*], all things, whether on earth or in heaven, making peace by the blood of his Cross. (Col 1:12–20; emphasis added)

To start with, let us set down a chronological note: "At the Passover in 30 A.D., a Galileean Jew was nailed to the cross in Jerusalem on account of messianic activities. About 25 years later, the former Pharisee Paul quotes . . . a hymn about this crucified person".[6] Paul sees the Galilean as the "first-born of all creation" who was before all else, "for in him all things were created, in heaven and on earth". The one who was crucified is a being equal to God and has truly cosmic significance. "In him all things hold together." He is not only the one who "holds the world together in its core", but he is also the one who brings universal reconciliation, "making peace

[6] M. Hengel, *Der Sohn Gottes* [The Son of God], 2nd. ed. (Tübingen, 1977), p. 9.

by the blood of his Cross" so that all things are "reconciled" "through him".

How did the little groups of Christians scattered across the world at that time come to see their founder, a few years after his death, in such a tremendous universal perspective—and similarly, to see themselves, the Church, as his body through which he accomplishes his work of universal reconciliation and union? Either these "messianic communities" were quite "nuts", sectarians blind to all reality, with ideas that were sheer fantasy, or they were the bearers of a great light, of a vision that offered entirely new insights into the reality of this world. This great vision unites the Cross and the cosmos, and even today it holds within it an incomparable potential for opening up new meanings. It is an exciting and fascinating vision, the power of which has not been lessened by the enormous progress made in natural science.

The hymn is divided into two symmetrical strophes. The first (verses 12–17) concerns creation, the second (verses 18–20) redeemed creation, the Church. There is in each case a threefold designation of Christ's function: "in him", "through him" and "to him". Both the cosmos and the Church are similarly planned and founded entirely *in him*, realized *through him*, and from the beginning are directed *to him*. To see the cosmos and the Church as united under the one head is an inspiring vision. In both strophes, Christ is seen in his role of origin and head, the leader comprehending the entirety of things. And "the entirety" here includes what is "visible" and what is "invisible", that is to say, "thrones or dominions or principalities or authorities", everything "on earth" and everything "in heaven".[7]

[7] On the passage as a whole, see the stimulating remarks of Cardinal Giacomo Biffi, *Gesù di Nazaret centro del cosmo e della storia* [Jesus of Nazareth: Center of the cosmos and of history] (Turin, 2000), pp. 135–52.

Parallel to this in many respects is the opening hymn to the Letter to the Ephesians, in which Christ is seen as the one through whom God realizes his plan "to unite all things in him, things in heaven and things on earth" (Eph 1:10). Similarly, the opening words of the Letter to the Hebrews are especially impressive:

> In many and various ways God spoke of old to our fathers by the prophets; but in these last days he has spoken to us by a Son, whom he appointed the heir of all things, through whom also he created the world. He reflects the glory of God and bears the very stamp of his nature, upholding the universe by his word of power. When he had made purification for sins, he sat down at the right hand of the Majesty on high, having become as much superior to the angels as the name he has obtained is more excellent than theirs. (Heb 1:1–4)

In this passage, too, we find the three dimensions that are articulated in the hymn of the Letter to the Colossians. The Son of God is the origin, the center, and the goal of the universe. Just as in the Letter to the Colossians, these three dimensions—the origin, the present, and the goal being aimed at—apply just as much to the created order as to the order of salvation ("he had made a purification for sins").

Easter as a New Creation

The prehistory of this vision is the vision of hope of the Jewish people, held fast over a long period through severe trials. No other people has accepted the created status of the world, and of man, at such a profound level, and carried it forward. No other people was so deeply convinced of the wisdom and rationality of the Creator's works. Yet equally, no other people took so realistic a view of the power of negative factors, of evil, of the wicked one, who frequently distort the world and history and threaten to destroy "God's undertaking". Only against the background of this coolly rational, yet hopeful, understanding of the world and of history, does it become possible to understand why such an "explosion" of hope could come about through Jesus Christ.

How was it possible only a few years after Jesus' death on the Cross, for the little communities of his disciples to sing a hymn like that in the Letter to the Colossians? There is only *one* credible answer: the Resurrection of Jesus. Easter morning changed everything. It gave to the age-old images of hope from the prophets concerning a new world and a reconciled cosmos a sound and concrete basis. "The Lord is risen indeed"—this Easter cry of the terrified apostles is witness to the initial kindling of a new way of looking at the world. As if in an intensive and concentrated white light, all the colors of the spectrum were already present here, and were then able to fan out in the process of developing reflection upon what had occurred.

First of all, an answer appears to what is probably the most urgent human question: Why are there pain and death? These realities clearly argue against any Creator supposedly having made his work with a loving purpose and with a loving will. The experience of the early Church, however, was this: "Give thanks to the Father with joy! . . . He has delivered us from the dominion of darkness and transferred us to the kingdom of his beloved Son" (Col 1:12–13). With the Resurrection of Jesus, redemption from "the power of darkness" has been won. He is "the first-born from the dead" (Col 1:18), and is thereby the beginning of the new creation (cf. 2 Cor 5:17) That is why what Paul says about the consequences of the Resurrection takes the form of a hymn: "When the perishable puts on the imperishable, and the mortal puts on immortality, then shall come to pass the saying that is written: 'Death is swallowed up in victory.' 'O death, where is thy victory? O death, where is thy sting?' . . . But thanks be to God, who gives us the victory through our Lord Jesus Christ" (1 Cor 15:54–55, 57).

Out of the certainty of the Resurrection of Jesus there springs the early Church's certitude that the Risen One is the goal of the universe, that he is the One toward whom everything has been created. The Resurrection of Jesus is, so to speak, the ultimate guarantee of "negative entropy": things are not moving toward death, but life. The universe is not the blind interplay of meaningless forces, but a "progetto intelligente", an "intelligent plan", as Pope Benedict XVI has called it,[8] a project to which, so to speak, a "happy ending" has been assigned.

This hopeful Christian view, however, has no need either to deny what is negative or to explain it away as "so-called evil" (Konrad Lorenz). It does not separate hope from suffering but sees the cosmos and the Cross together. In this view, the Cross—as is beautifully clear in the Good Friday liturgy—is the tree of life, the wood upon which life has overcome death, and love is victorious over evil. The "intelligent plan of the cosmos", about which Pope Benedict speaks, is a plan of love.

Only Christ can offer us the certainty that this is not an illusion, a projection of human longings that can never be realized. Paul found in Christ the unshakeable certainty that his love is victorious. That is why the hymn in the Letter to the Colossians sees Christ as being to some extent the model of what the creation of man was aiming at. Christ "is the image of the invisible God", and each person is an "icon of Christ".[9] Hence, in my view, speculations such as

[8] *General Audience*, November 9, 2005.

[9] Cf. Biffi, *Gesù di Nazaret centro del cosmo e della storia*, p. 143.

those of Hoimar von Ditfurth, which want to look at contemporary man "merely" as a "transitional stage" of evolution on the way toward a "superman", are superfluous. In Christ, we have been presented with the true goal of the "evolution" of man: being formed in his likeness is what complete, fulfilled, and successful humanity can achieve. We do not need to look forward to a utopian, "post-human" form of *homo sapiens*, but to the gracious and perfect form of man, as we may glimpse it in the saints, and as it is wholly realized in Christ.

If Christ is the goal of the "project", then it has also been planned *in him* from the beginning, and is being realized *through him*. In the prologue to his Gospel, John sees Christ as the *Logos*: "In the beginning was the Logos, and the Logos was with God, and the Logos was God" (Jn 1:1). What does "*Logos*" mean? It does mean "word", certainly—but also "meaning", "reason", and "essential determining factor". "All things were made through the Logos, and without the Logos was not anything made that was made" (Jn 1:3). The great tradition of Christian thought has understood that there exists an inner connection between the *Logos* and the *logoi* of things. The "logos" of a thing is what makes it what it is, what determines it in its inmost being, what decides its nature. If each creature thus has its own "logos", then the trace of the Creator-Logos is found in each one. That is what is meant by the doctrine of the *vestigia dei*, the "signs of God", in creatures. God's traces in creation are the signs of the Logos, of Christ, in whom and through whom and toward whom everything was created.

That is why Christ the Logos does not come as a stranger into this world either. Rather, as John tells us, "He came to his own home" (Jn 1:11). "*Alles ist dein Eigentum*" ["Everything belongs to thee"] we sing in the third verse of the German version of the *Te Deum*. John further says concerning Christ the Logos, "In him was life, and the life was the light of men .... The true light that enlightens every man was coming into the world. He was in the world, and the world was made through him, yet the world knew him not" (Jn 1:4, 9–10). But this means that the *Logos* is the Word of creation that re-echoes in the inmost part of every creature; it is the creative rationality that gives meaning to all things and that grants them their nature and their effective action. And the *Logos* is the light that is active in our reason, illuminating it and making it clear, so that it can penetrate the realm of creatures and is able to recognize in things the signs of the Creator. The Logos of God is at work both as the trace of the Creator in the creatures and as light in human reason, so that this reason may become aware of the signs of the Creator and may recognize them.

Saint Thomas Aquinas says that (created) things are located between two "reasons" (*res inter duos intellectus constituta*): between the divine reason that created them, and the human reason that is able to recognize them. Things are capable of being perceived because they are rational.[10] We are able to perceive them, because in our reason we have a share in God's reason.

Now, someone may counter, "What do these high theological ideas mean for the natural scientist, who goes about his daily work with cool rationality, taking painstaking steps forward in our knowledge—mainly small ones, but sometimes big ones?"

The question of God rarely arises as a purely philosophical question. Simply following a purely inductive path, starting from one's observations of the world, one will hardly discover all that we have heard in the early Church's hymns to Christ. We of the Catholic faith do indeed firmly hold

[10] *Quaestiones disputatae de veritate* I, 2.

that we can become aware that "what can be known about God is plain . . . his invisible nature . . . clearly perceived in the things that have been made" (Rom 1:19–20). Yet the "surpassing worth of knowing Christ", about which Paul talks (Phil 3:8), is purely a gift of grace.

Knowing Christ means, above all, being known by him. Meeting him means being found by him. "In him are hid all the treasures of wisdom and knowledge," says Paul in the Letter to the Colossians (Col 2:3). Finding Christ means finding the one in whom all things hold together, who sustains it all through the power of his word. Nothing can substitute for knowing Christ, for being united with him, for having a share in his suffering and in his risen life. The difficulties of actual practical work are not thereby lessened, but a marvelous horizon of significance opens up. The knowledge of Christ strengthens the often weak and flickering light of reason, and it grants us that straightforwardness of living through which reason, too, becomes clear and bright. Finally, knowledge of Christ grants to us that hope which never disappoints us. This is the hope that all the fascinating beauties of creation, all its wonderful complexities, the most finely adjusted interplay of forces and elements—the hope that all of this is not an empty game over the inevitable abyss of death, but rather the premonition of a world to come in which death no longer has the last word. And Christ himself, in his whole existence, with his life and his suffering, is the sole really sustainable response to the tormenting question of evil in the world. Only in him can the restless heart find peace. He alone was able to give to suffering, to failure, disaster and death, an unsuspected value. What seems according to reason to be valueless, in the light of Christ acquires a significance and begins to shine out in the credible testimony of men.

Teilhard de Chardin—Witness to Christ

Hardly anyone else has tried to bring together the knowledge of Christ and the idea of evolution as the scientist (paleontologist) and theologian Fr. Pierre Teilhard de Chardin, S.J., has done. His fascinating vision has remained controversial, and yet for many it has represented a great hope, the hope that faith in Christ and a scientific approach to the world can be brought together "under one head", under Christ the "evolutor".

Teilhard understands the universe as being in a great upward movement to ever greater complexity and inwardness, from matter to life to mind. It is a movement with a goal (and therein Teilhard is differentiated from those who assume that evolution has no direction), leading from geogenesis to biogenesis and then to psychogenesis. This upward movement is completed, however, when "Christogenesis" comes forth from cosmogenesis. In this ascent, evolution ceases to be passively accepted with the appearance of man, and the stage of auto evolution has been attained. This, in turn, reaches its climax with the appearance of Christ. He becomes the visible center of evolution as well as its goal, the "omega-point". The incarnate Logos, who appears in visible form at a certain point along the evolutionary axis, had previously been the invisible "motor of evolution". Christ, as head of the cosmic body, fulfills everything, guides everything, and perfects everything. "The entire universe is ipso facto stamped with his character, determined by his choice, and animated by his form." According to Teilhard, Christ becomes the energy of the cosmos itself. For through the Incarnation, God himself has become "immersed" in matter, and within it and from the midst of it, he effects "the leading and planning of what we nowadays call 'evolution'". The Incarnation brings about a kind of "Christification" of the cosmos.

Teilhard de Chardin also sees the Cross of Christ within this perspective. It becomes the impulse for the overcoming of what is lacking in cosmic development. Finally, through his Resurrection Christ is freed from all limitation to his power and to the effectiveness of his activity, and he is able to guide the cosmic development toward the omega point, the world's ultimate "amorization" (turning into love), which will be perfected in the parousia, the return of Christ.[11]

These brief references to Teilhard cannot do justice to his efforts. The fascination which Teilhard de Chardin exercised for an entire generation stemmed from his radical manner of looking at science and Christian faith together. This unity of vision, in which he intended to unite natural science and Christian faith, was of course also problematical. Critics have shown that he could not do complete justice to both sides. His vision of evolution as an upward movement that ceaselessly produces higher and ever higher forms is more of a philosophical speculation than a scientific theory. On the other hand, his "naturalization" of Christ as the driving force of evolution inevitably ran up against contradiction in theological terms. Despite the criticisms from both sides, many people have come to feel his concerns and have valued them. Above all, the way he was fascinated by Christ is impressive. His love for Christ made him into a kind of "mystic of evolution". In this he is far removed from the materialistic concepts of the "evolutionism" that is widespread nowadays. For our subject, it is important that Teilhard de Chardin dared a venture that was at the same time

[11] All quotations are translated from Leo Scheffczyk, "Die Christogenese Teilhards de Chardin und der kosmische Christus bei Paulus" [Teilhard de Chardin's Christogenesis and the cosmic Christ in the writings of Paul], in idem, *Schwerpunkte des Glaubens: Gesammelte Schriften zur Theologie* [Focal points for faith] (Einsiedeln, 1977), pp. 249–79.

full of risks and yet necessary. He incorporated the way that the Christian faith viewed the Incarnation of God in Jesus Christ as an inspiring vision into his research and his thought as a natural scientist. Conversely, he was constantly opening up his activity as a scientific researcher toward the great horizon which had been unlocked for him by his Christian faith.

It is true that faith and science should be distinguished from each other. Yet it is also true that they ought not to be separated. Science has need of the broad horizon of faith. Through his work, Teilhard de Chardin helped many scientists to overcome the prejudice that faith cramps science. Faith in Jesus Christ, in whom all the treasures of wisdom and knowledge are hidden (see Col 2:3), deprives science neither of its freedom nor its zest, neither of its honesty nor its enthusiasm—on the contrary, indeed, it further strengthens it.

VIII.

"Subdue the Earth" (Genesis 1:28) Responsibility for Creation

Hardly any sentence in the Bible seems so problematical today and is so often rejected as the famous command, "Subdue the earth!" The whole world today, so it is said, is suffering the consequences of this divine command to mankind. The well-known German writer Carl Amery gave his book *Das Ende der Vorsehung* [The end of providence] the characteristic subtitle *Die gnadenlosen Folgen des Christentums* [The merciless consequences of Christianity].[1]

Christianity is in the dock again, this time not as the great obstacle to progress, the conservative "stopper" that always fearfully criticizes and rejects everything new, but as the evil protagonist of progress, justifying the total plundering of nature on a biblical basis. Is this biblical command not the absolute contrary to all that moves so many people world-wide today, under the heading of protecting the environment? In the Book of Genesis, we read:

> God blessed them [male and female], and God said to them, "Be fruitful and multiply, and fill the earth and subdue it; and have dominion over the fish of the sea and over the birds of the air and over every living thing that moves upon the earth" (Gen 1:28).

This passage sounds like mockery in the face of the warning the Club of Rome pronounced to a shocked Western world at the beginning of the seventies: a vision of overpopulation like something out of a horror-film; an ideology of growth whose dangers were suddenly being listed; an optimism centered on progress that was beginning to turn to panic. With the rapid globalization since then, the reasons for concern have not decreased.

[1] Reinbeck, 1972.

Is the Judaeo-Christian tradition, which urges us to achieve growth and dominion over the world as the first of all the God-given commandments handed down in the Bible, the original cause, so to speak, that has provoked the ecological catastrophe toward which we are moving—or in which, indeed, in the view of many, we are already irreversibly engaged? Here we have to ask more precisely what the biblical commandment implies.

Man's Dominion in the World

Two-and-a-half millennia years ago, of course, when the text of the Book of Genesis reached its final form, overpopulation was "still so far beyond people's range of vision that [the authors] had no reason to issue warnings about it."[2] The concern for having enough descendants and for an environment that would support life and was liveable was foremost.

Genesis 1:28 contains a couple of presuppositions that are being questioned today in two ways. The first presupposition for the command to subdue the earth is the creation of man as male and female. On this point, it says:

> Then God said, "Let us make man in our image, after our likeness; and let them have dominion over the fish of the sea, and over the birds of the air, and over the cattle, and over every creeping thing that creeps upon the earth." So God created man in his own image, in the image of God he created him; male and female he created them. (Gen 1:26–27)

Man is seen as a society from the very start: "Let *them* have dominion ..." Between them, as the family of humanity,

[2] J. Scharbert, *Genesis I–II*, 4th ed. (Würzburg, 1997), p. 46.

they are assigned the task of being God's deputies. As beings created "in his image", they share in God's dominion over the world. They have been entrusted by him with the "stewardship of the world".[3] But they have received this stewardship not as an anonymous "humanity", but in the concrete form in which the family of man actually exists, as "male and female" (Gen 1:27b).

Secondly, we have to clarify what we mean by the word "dominion". "The manner of human stewardship of the world is unconditional superiority", is the information biblical studies furnish on this point. The Hebrew word means "always ... an action in which man applies his power to make use of something. Thus humanity, in the image of God, is furnished with capabilities and is authorized to dispose of the world."[4] The eighth Psalm, praising the Creator of man, says clearly and unequivocally, "Thou hast given him dominion over the works of thy hands; thou hast put all things under his feet" (Ps 8:6).

Every living thing on earth and in the sea and in the air is under man's dominion, yet there are two important qualifications to this. First, all men—not just a few—have in common the task God has given them to be stewards of creation, to shape it and to have it at their disposal—to be entrusted with it. From this is derived the fundamental principle of Catholic social teaching: "The goods of creation are destined for the whole human race."[5] Second, man is entrusted with dominion over non-human creation not dominion over other men: "Only man himself is not to be subjected to domination".[6]

[3] H. W. Wolff, *Anthropologie des Alten Testaments* [Anthropology of the Old Testament], 2nd ed. (Munich, 1974), p. 236.

[4] Ibid., p. 239.

[5] CCC 2402.

[6] See Gen 9:6; Wolff, *Anthropologie des Alten Testaments*, p. 240.

Listening to Creation

What kind of dominion is man to hold over the creation that is subject to him? Cardinal Ratzinger, now Pope Benedict XVI, in his homilies on the theme of creation delivered at Munich in 1981, expressed it quite clearly: "The Creator's directive to humankind means that it is supposed to look after the world as God's creation, and to do so in accordance with the rhythm and logic of creation."[7]

The Creator himself gives us the yardstick for this dominion he has set for us men to exercise in his creation—through its language, its rhythms, its meaning, and its logic. Thus, this concerns the question of whether we are in the position, as "stewards of creation", to be aware of the language of creation and to give it our attention. Man's commission as steward of creation means that he understands how to use creation "for what it is capable of and for what it is called to, but not what goes against it".[8]

But does creation have a language we can hear? Is there an order in creation to which we have to pay attention? Are there recognizable indications of the Creator in creation that teach us in what sense we may be aware of something like a "responsibility for creation"? This is precisely the point at issue. People often question whether there are indications given through the being of creation, out of which it follows that we have some responsibility. Evolutionism, as an ideology (to be distinguished from the scientific theory of evolution or descent), denies any such possibility. I mention two spheres in which there are also reservations about an order in creation, in the way many people feel nowadays.

[7] Joseph Ratzinger, *"In the Beginning . . .": A Catholic Understanding of the Story of Creation and the Fall* (Huntington, Ind.: Our Sunday Visitor, 1990), p. 48.

[8] Ibid.

The first is the intimation of the Creator of man that the latter has been created "as male and female", and was intended thus: it is becoming difficult to recognize—and this is more or less at the prompting of the "spirit of the age"—that being a man or being a woman is not just a culturally accepted notion of an arbitrary nature, nor a chance genetic product, but is in the first instance a primary characteristic of being. There is no question that the play of genetic factors and cultural influences exert an influence here. Yet they do not constitute what it is to be a man or a woman. They may be able to influence the way this is shaped in practice, yet always within the given fact that they are shaping the actual form of being a man or being a woman. Yet if there is widespread failure to understand the primary fact, that is even more true for the conclusions drawn from this in terms of morality—for instance, in the homosexual question. Such a view of our responsibility for creation is all too completely in contradiction with another view of dominion of the world, which is widely prevalent.

A second example of responsibility for creation, understood in a Christian sense, is animal protection. I quote as an example the three brief articles relating to this in the *Catechism of the Catholic Church*. They are somewhat dry, stylistically, but the moral pointers are entirely in keeping with reason and with belief in creation. The protests that rained down against this article bear witness to a failure (often a profound failure) to understand the commission to man, in the Bible, to exercise dominion. The *Catechism* states:

> *Animals* are God's creatures. He surrounds them with his providential care. By their mere existence they bless him and give him glory. Thus men owe them kindness. We should recall the gentleness with which saints like St. Francis of Assisi or St. Philip Neri treated animals.

> God entrusted animals to the stewardship of those whom he created in his own image. Hence it is legitimate to use animals for food and clothing. They may be domesticated to help man in his work and leisure. Medical and scientific experimentation on animals, if it remains within reasonable limits, is a morally acceptable practice since it contributes to caring for or saving human lives.
>
> It is contrary to human dignity to cause animals to suffer or die needlessly. It is likewise unworthy to spend money on them that should as a priority go to the relief of human misery. One can love animals; one should not direct to them the affection due only to persons.[9]

The first of these three articles meets with no criticism, apart from the fact that the term "kindness" towards animals seemed to many people somewhat insufficient. Above all, however, the final sentence of the third article gave rise to violent controversy: one may "love" animals, it is said, but not direct to them the "affection" due only to persons. Yet it ought to be obvious that the order of creation is somehow distorted whenever animals are provided with luxuries while at the same time men are deprived of the most essential things. That does not mean that love for animals is something bad. There is simply an *ordo caritatis*, an order in loving, which is in accordance with creation and which becomes immediately obvious when we are confronted with the case of parents who give everything possible to their pet dog but allow their children to starve. Or think of Rudolf Höß, the camp commandant at Auschwitz, who loved animals and was at the same time a cold-blooded mass murderer.

People have already pointed out, a good many times, the painful fact that the animal protection lobby is given a better hearing in politics today than those who try to protect

[9] CCC 2416–18.

unborn children. The order of creation should provide a clear ranking of imperatives here. Yet even the notion of such a rank, a prioritizing, is understood by many people as if giving priority to the protection of human life before birth were an affront to quite justified efforts to protect animals.

But let us look at the *positive side* of the command concerning creation. The Lutheran exegete Ludwig Köhler tried to formulate in an easily comprehensible way what Genesis 1:28 means in the context of the early development of human culture, as well as in contemporary cultural terms:

> The command to take dominion in creation is the *command to establish culture*. It is directed to all men; it encompasses all eras; there is no human activity that is not subject to it. The man who found himself and his family on the unprotected steppes at the mercy of an icy wind, and who first laid a few stones on top of one another and thus invented the wall, the basis of all architecture, was fulfilling this command. The woman who first bored a hole in a hard thorn or a fish bone and drew a length of animal sinew through it, so as to be able to join together a couple of scraps of pelt, and who thus invented the needle, sewing, and the beginning of all clothing manufacture, was fulfilling this command. To this day, every time someone gives instructions to a child, every kind of school, every book, all technology, research and science and teaching along with their methods, instruments and institutions—all that does nothing but fulfill this commandment. The whole of history and all human strivings come under the heading of following this command.
>
> This is the objective side. There is also a subjective side. Each man—and this lies in his nature, never to be lost—somehow has to cope with life. Everyone has to try to cope, inwardly, with whatever happens to him, whether it be a speck of dust in his eye that hurts him, or a flood that threatens his

> life and the lives of those dear to him—nothing is too great, and nothing too small. . . . A man's nature is known by the way he deals with things inwardly.[10]

"Subdue the earth" is thus man's command to establish culture from the very beginning. This work is not yet finished, but remains for every age, for each generation, an enduring task. It is ever anew under threat from perversions of the command concerning creation, especially when we start to forget that this is a commission, a task, and not an arbitrary and self-centered will to dominate—not a license for the unlimited plundering of creation.

Right and Wrong Understandings of Dominion

It is precisely this danger that is involved in the contemporary situation in which many people view the command to "subdue the earth" as the great danger to the ecological and human future of the world. Cardinal Ratzinger, in his homilies on creation, talked about a "change of paradigm in our understanding of the commission to exercise dominion" delivered to man:

> How did the mentality of power and activity, which threatens us all today, ever come to be? One of the first indications of a new way of looking at things appeared about the time of the Renaissance with Galileo, when he said that if nature did not voluntarily answer our questions but hid her secrets from us, then we would submit her to torture and in a wracking inquisition extract the answers from her that she would otherwise not give. The construction of the instruments of the natural

[10] Ludwig Köhler, *Der Hebräische Mensch: Eine Skizze* [Hebrew man: An outline] (Tübingen, 1953), pp. 112–13, cited by Wolff, *Anthropologie des Alten Testaments*, pp. 240–41.

> sciences was for him as it were a readying of this torture, whereby the human person, despot-like, gets the answer that he wants to have from the accused.[11]

At that time, a new kind of knowledge was being sought—not what things are, what constitutes their "nature", or, to put it another way, what their "logos" is, the divine idea that is being expressed in them—but rather what we can make out of them for ourselves. This approach to reality is called "power knowledge". René Descartes exercised a decisive influence on modern scientific thought and described this new kind of knowledge as follows:

> It is possible to attain to a knowledge which can be very useful in this life. And instead of that kind of speculative philosophy that is taught in the schools, we can discover a practical philosophy by which, knowing the power of fire, water, air, the stars, the heavens and all the other bodies which surround us, and knowing the way they work. . . , we will be able to make use of them . . . in all the various applications for which they are suited; and thus we will make ourselves, as it were, masters and owners of nature.[12]

In order to establish this kind of "power knowledge", the idea of a Creator had to be set aside. It had to be made possible as a "hypothesis", to dispense with him. It was a matter of eliminating any "language of creation", and thereby any message from the Creator, so that sheer utility, with no limitations or handicaps, becomes the model that determines everything. The question regarding a Creator had to be declared meaningless, as did the question of his commission to us in creation. In its place a different basic commission for man had to be formulated: "This is the source

[11] Ratzinger, *"In the Beginning . . ."*, p. 49.

[12] Discours de la méthode 6:2 (Paris: Garnier-Flammarion, 1966), p. 84.

of the change in humanity's fundamental directive vis-à-vis the world; it was at this point that progress became the real truth and matter became the material out of which human beings would create a world that was worth being lived in."[13]

The two ideologies that this "power knowledge", dissociated from the Creator and the creation, brought into a largely dominant position during the nineteenth and twentieth centuries were Marxism and Darwinism (to be distinguished, once again as an ideological trend, from Darwin's scientific theory). The young Karl Marx drew up the following basic outline:

> A *being* cannot call itself independent until it can stand on its own feet; and it stands on its own feet when it owes its existence to itself. A man who lives by the grace of someone else regards himself as a dependent being. And yet I live entirely by someone else's grace if I owe to him not only the sustenance of my life, but if besides that he has *created* my life; if he is the source of my life—and my life necessarily has such a basis outside itself, if it is not my own creation. *Creation* is thus a concept that is difficult to drive out of the consciousness of ordinary people.[14]

Not to owe one's existence to anyone else is certainly the presupposition for empowering oneself to be one's own Creator. It is certainly no accident that Friedrich Engels, after reading Darwin's *Origin of Species*, wrote enthusiastically to Marx that here they had found the scientific basis for their theory.

[13] Ratzinger, *"In the Beginning . . ."*, p. 50.

[14] Karl Marx, translated from *Nationalökonomie und Philosophie. Frühschriften* [National economy and philosophy: early writings] (Stuttgart, 1953), p. 246.

As a decisive negation of belief in creation, (popularized) Darwinism was not merely propagated by Marxism right to the end of Communism; something similar also happened in "right-wing" ideological materialism as it was widely publicized in the German-speaking area, especially by Ernst Haeckel. In this case, it was combined with racial theories and "social Darwinism", with its cult of the strongest, finding its most horrific expression in National Socialism.[15]

This ideology was vanquished, and even the Marxist utopia has perished, wrecked upon a reality that would not conform to its ideology. It collapsed because it would not admit that there is any such thing as human nature, or an order in creation that we cannot contravene without being punished. This became especially clear in the realm of economics, but above all in the erroneous conception of man that underlay the ideology.

The era of ideologies is at an end. However, the paradigm of "power knowledge", the ideology of the domination of nature as mere matter for our scientific/technical/economic will to power to work on—a domination dissociated from belief in creation, and the responsibility for creation that is founded on it—has not yet perished. Now, as in the past, our culture is dominated by a way of thinking that sees the world as a product of chance and necessity rather than as the utterance and the challenge of the Creator. Now, as in the past, we find it extremely difficult to take into account in the great ethical questions of our time any suggestion from the natural order and its Creator. Now, as in the past, the idea of utility is largely dominant within the ethical debates about whether what we can do is permissible.

[15] Carl Amery, *Hitler als Vorläufer* [Hitler as precursor] (Munich, 2002), pp. 22–33.

Behind the debate about creation and evolution, there does not stand the frequently popularized alternative of "a return to the Middle Ages through ecclesiastical, anti-scientific dogmatism" versus "scientific progress looking to the future". What is really at stake is the great question, decisive for our future, of basic ethical guidelines capable of underpinning a high-technology civilization and its capabilities, which threaten our very existence. It is a matter of whether science can be combined with a sustainable responsibility for creation. For science, understood correctly, is a practical application of the first divine commandment to men: "Subdue the earth." That is why responsibility for creation must also be a challenge to the self-understanding of science.

Responsibility for Creation and Science

The alternative "evolution or creationism" is too simplistic. "Creation and evolution" is another, substantially more serious challenge. Here, in my view, is where the question of the great moral choices is being decided today. Is man simply a material that can be used one way or another? Is human dignity an unconditional initial datum, which is part of the order of creation and never merely represents a concession? Can this view of human dignity, upon which our democracies, our systems of a just society, and our understanding of social solidarity are based, continue to have no effect on the scientific picture of man? And can the various attempts to broaden the theory of evolution so as to apply it to human knowledge, human thinking, and the human will—to human society, economy, and ethics—do justice to the unconditional demands of the dignity of man?

The great Jewish philosopher Hans Jonas, in his late work *The Imperative of Responsibility*, to which he gives the subtitle *In Search of an Ethics for the Technological Age*, addresses the decisive question of whether a "should" follows from an "is", whether an imperative follows from existence.[16] In theological terms, he is asking whether creation has a language that can give us any counsel. If everything is merely the product of chance and necessity, no counsel, no positive suggestion, can stem from creation, nor does it have any law of its own. If, however, it has its own being, a being intended by the Creator, then—and only then—can there also be a responsibility toward it. This "responsibility for creation" only exists if an imperative is directed to us from the existence of creation. Many people cast doubt, in theoretical terms, on whether there is such a connection between existence and imperative. Hans Jonas employs a simple example that he calls the "primeval object of responsibility": the child, "the newborn child, whose mere breathing incontrovertibly directs an imperative to its environment: namely, to accept him. Just look, and you will know".[17]

To resist this call seems inhuman to us. We are helped in this, of course, by instincts and feelings towards a small baby that are "programmed" into man. Behavioral research has accepted the evolutionary "natural history" of this model of behavior. These "crude caring instincts", the roots of which lie far back in the animal world, are relationships of absolute necessity. They cannot explain, however, the character of unconditional moral imperative attaching to care for a newborn human child. The essential difference between

[16] Hans Jonas, *The Imperative of Responsibility: In Search of an Ethics for the Technological Age*, trans. Hans Jonas and David Herr (Chicago, 1984); German original: *Das Prinzip Verantwortung: Versuch einer Ethik für die technologische Zivilisation* (Frankfurt: Insel, 1979; Surkhamp Taschenbuch, 1984).

[17] Jonas, *Prinzip Verantwortung*, p. 235.

the pre-programmed crude animal caring behavior and human care for a newborn human child is found in the genuine sense of responsibility that is peculiar to man alone. That is why he alone cannot fail in his responsibility without incurring guilt. This distinction may, indeed, be denied at a theoretical level. Yet it is because there is genuine responsibility that a failure in our duty to help in an emergency renders us liable to punishment.

What is there in creation more sublime, more precious, than a new human child? What is in greater danger, nowadays, than an unborn child? It is hard to understand how one's commitment to environmental protection is not turned as a priority to protecting children. For nothing demands our respect for creation more than care for its most precious possession: the child that has been vouchsafed as a gift to this world in order to know it, respect it, and cultivate it.

IX.

Summary and Prospect

At the symposium entitled "Christian Faith and the theory of Evolution", in Rome in 1985, when I was still professor at Fribourg, I was able to contribute a paper of my own, to which I gave the title "The Catechesis of Creation and the Theory of Evolution: from Truce to Constructive Conflict".[1] The *Frankfurter Allgemeine Zeitung* reviewed the symposium volume under the title "The Truce is at an End".[2] Even today, the question of the relation between the theory of evolution and the belief in creation is discussed intensively once more. I can only feel pleased about this. Nothing is more damaging, to both sides, than when nothing is happening. "What affects everyone should be dealt with by everyone", runs an old adage.

Nothing affects all of us more than the primeval questions of man about where he comes from and where he is going, about the meaning of his life. Every person must ask these questions for himself if his life is to be a truly human one. Is there no answer to be found? Where can the answer come from? How can we tell whether an answer helps us to find the right direction in our own life and in society? Where, for instance, shall we obtain the criteria to establish an ethically responsible standpoint in the hotly debated question of how embryonic stem cells are obtained?

The key question upon which everything depends is this: Are the world in which we are living and our life in it meaning-full? A thing is only meaningful, if it has an end,

[1] Christoph Schönborn, "Schöpfungskatechese und Evolutionstheorie. Vom Burgfrieden zum konstruktiven Konflikt", in R. Spaemann et al., eds., *Evolutionismus und Christentum* [Evolutionism and Christianity] (Weinheim, 1986), pp. 91–116.

[2] "Der Burgfrieden ist zu Ende", *Frankfurter Allgemeine Zeitung*, February 13, 1987, p. 11.

a purpose. Without rationality there is no direction, no plan, no *design*. Cardinal Ratzinger has commented on this point:

> The Christian picture of the world is this, that the world in its details is the product of a long process of evolution but that at the most profound level it comes from the *Logos*. Thus it carries rationality within itself.[3]

In everything I have said thus far, I have constantly been concerned with the intermediary role of reason. The conflict is frequently reduced to one between science and religion. Yet it is decisively concerned with the "linking element" in both, with reason. It is reason that recognizes direction toward an end, plan, intention, design, and purpose in nature, and it does so in ever greater measure. The more we are able to know, the more complete and detailed our knowledge of the processes of life, the greater—in my view—our wonder and admiration should become. Conversely, it becomes all the more irrational to derive all of this (as I said in the *New York Times*) from an "unguided, unplanned process of random variations and natural selection".

Stanley Jaki, a Benedictine and specialist in the history of science, once said that it is remarkable enough that there are "Darwinists" who devote their entire scientific careers to demonstrating that there is no purpose.[4] Has the theory of evolution disproved the existence of finality, of purposefulness? Fr. Jaki, in a sharply ironical tone, says:

> The greatest difficulty for this assertion is as follows: a supposedly aimless process of evolution finished by producing a being,

[3] Joseph Ratzinger, *God and the World* (San Francisco: Ignatius Press, 2002), p. 139.

[4] Stanley L. Jaki, *The Road of Science and the Ways to God* (Edinburgh, 1978), p. 281.

> man, who does everything with a purpose. Even evolutionists have a purpose in denying any purpose: their purpose is to promote materialism—which is certainly no kind of science, but a counter-metaphysics.[5]

Omne agens agit propter finem (Every activity is done for a purpose), so runs a basic principle of classic metaphysics in Aristotle, Thomas Aquinas,[6] and many others. Once again, it is legitimate, and may be justified as a method, to exclude the question "Why?", the search for finality, from a certain way of regarding nature. But it is not legitimate—indeed it is irrational—to conclude from that, that there is no finality.

The aggressive way in which many oppose the group of American scientists who are devoting themselves to investigating "intelligent design" does not have much to do with science. One may criticize their methodological approach. Yet the question as to the origins of the obvious "intelligent design" in living things is an entirely legitimate one; indeed it is a question bound up with man and his human reason. An answer to this question is not to be expected from research working along strictly scientific methodological lines, yet the question is set before man as a being who questions things, wonders at them, and thinks about them.

Where, then, is the answer to be found? I would like to put forward a theory about this, which may be illustrated by a metaphor that I found in a book by Joachim Illies: the image of two ladders, Darwin's ladder and Jacob's ladder.[7] This image is meant to symbolize the ascending movement

[5] Stanley L. Jaki, "Non-Darwinian Darwinism", in R. Pascual, ed., *L'Evoluzione: Crocevia di scienza, filosofia e teologia* [Evolution: crossroads of science, philosophy and theology] (Rome, 2005), p. 44.

[6] Cf. *Summa Theologiae* II–II, q. 1 a. 2; I, q. 44 a. 4.

[7] Joachim Illies, *Schöpfung oder Evolution. Ein Naturwissenschaftler zur Menschenwerdung* [Creation or evolution: a natural scientist writes on the Incarnation] (Zurich, 1979), p. 104.

of evolution and the movement of the Creator Spirit coming down from God. These are two movements in two different directions, which offer something like an overall view only when both are seen together. The way that both movements have their center in Christ, and their meaning and their inner goal in him (see chapter 7), is a theme we will take up again at the end.

Two Ways of Looking at Things—Two Stories

In the early seventies, Konrad Lorenz the behavioral scientist and Nobel prize winner, and Viktor Emil Frankl, the pupil of Sigmund Freud, concentration camp survivor and founder of "logotherapy", held a conversation that Frankl recounted in a lecture as follows:

> If so many people say that the whole of evolution is nothing but chance, and it results from random occurrences that have no significant interconnection, like mutations, and thus there is no kind of teleology (or direction towards a goal), then we have to ask, after all, whether this biological plane is necessarily the only plane on which we may be permitted to see any reality. Is it not possible, then, that the plane of biology is just *one* way of looking at things and that perhaps another plane—a "vertical" one in relation to this—might exist? And might it not be that on this other plane, in this system of coordinates, an *idea* may perfectly well exist, a line linking those points that on the horizontal plane seem to be unconnected (indeed, random)? Seen in this way (by taking account of a *vertical* plane), teleology (a combination of meaning and direction toward a goal) might lie even behind mutations.
>
> I do not expect Konrad Lorenz to *plunge into* this realm and then to say, "Yes, certainly, this orientation toward a goal, this significance, *does* exist." But I do expect him not to insist that

> the way of looking at things on the purely horizontal plane is necessarily the only one that exists. At that time I said to Konrad Lorenz, "You know, if you would simply admit that it is fundamentally possible that, on another plane from the biological one, a teleology, an association with meaning, a significant orientation, does exist, then you will have deserved a second Nobel prize: a Nobel prize for wisdom. For wisdom is science, plus the knowledge of one's own limitations."[8]

Of course, these two ways of looking at things are often considered de facto to be mutually exclusive, as actively opposed to each other. In the United States, the disputes even take place in the courtroom with enormous media attention. Perhaps it will help us to see things more clearly here if we talk not just about two ways of looking at things, but of two stories that have often been told as if they were in mutually exclusive competition with one another.

For centuries, the story of creation was told according to the creation account as presented in the Bible. Men understood themselves as part of a long history that began with Adam and Eve and was founded upon the six-day work of the Creator. The world was created about six thousand years ago, and even today the Jewish people reckon the years "from the beginning of the world". With the discovery of the vastness of the universe and the age of the earth, it is not surprising that this story has encountered ever stronger competition. The "scientific" story of the world has increasingly replaced the biblical one, turning it into a mythical story. It was only with the rise of the Darwinian history of life, however, that such sharp competition developed. Darwinism became an alternative story of creation, which of

[8] Viktor Frankl, *Altes Ethos—Neues Tabu* [Old ethos—new taboo], 1974; cited by Peter Blank, *Alles Zufall? Naive Fragen zur Evolution* [Is everything random? Naive questions about evolution] (Augsburg, 2006), pp. 75–76.

course no longer needed a Creator; and in comparison with the biblical story, the latter bore the enormously attractive, all-illuminating halo of science.

We have to realize clearly that everywhere today evolution is recounted as the valid history. As a form of history, it is dominant in school books, the media, and public debate as well as in advertisements, caricatures, and so on. And it is presented as claiming to tell us how things really happened. What is left to the biblical story is at best the narrow freedom of saying something about the meaning of human life. For anyone who wants a scientific picture of the world then the story of the world as recounted in accordance with an evolutionary model is the real, true story.

Yet both stories continue to be told. The *biblical* one is told in worship (at the Easter vigil, for instance), in religious education (sometimes with acute discomfort, and a great deal of watering down), or in the concert hall, whenever Haydn's *Creation* is performed. Everywhere else the "scientific" story is told—it is indeed often taken up in religious education precisely because "science" teaches it. The question is inevitable (and young people especially ask themselves), "Which of the two stories is true?" The answer is usually quite clear: the scientific story! How could it be otherwise? It is indeed represented as having been "long established by science".

How are we to explain, then, that since the publication of Darwin's *Origin of Species* the scientific debate has raged unchecked? All along, the "Darwinian story" has been extensively questioned by people who cannot be suspected of being "fundamentalists". So many questions still remain open that we continually have to be surprised at the self-assurance and emphasis with which the "Darwinist version" is recounted.

Strictly scientific research into evolution, which can report its progress step-by-step so that others can do the same, is a most respectable area of research. However, the extended application of evolution to all spheres of reality under the motto, "everything is evolution", no longer has any scientific basis. Here we enter into the realm of a worldview, if not of an ideology. The "Darwinian version" of things influences not only the way we conceive the origin of life and its development. It has also influenced our life in society and attitudes to the main moral questions in bioethics, in education, and in science—and continues to do so.

The correct path to follow is not to choose between the "Darwinian story" and "creationism", as people like to suggest, but a coexistence of "Darwin's ladder" and "Jacob's ladder". There is a great deal to support the view that life has developed through a long process, a gradual ascent from the most simple beginnings up to the complexity of man. It is marvelous to penetrate ever more deeply into the common building blocks of life and thereby into the way that all life is related. It is not only unnecessary, however, but contrary to reason, to view this grandiose path of life up to man as being an exclusively random process. When an astronomer, who is also a priest and theologian, even has the presumption to say that God himself could not know for certain that man would be the product of evolution,[9] then nonsense has taken over completely.

The alternative to a merely random process is not complete determinism, but the "limitation" of the creatures' own autonomous activity and of the divine creative Spirit, which sustains this activity and makes it possible. In the beginning was the Word, not random chance. Chance, in

[9] For example, Fr. George V. Coyne, S.J., in *Der Spiegel*, no. 52, December 22, 2000.

the sense of something unplanned happening, does exist, but it is not the great creative principle that ideological Darwinism would like to make of it.

To conclude, under brief headings we will mention three spheres in which it is becoming particularly clear today that a "world-view" constructed purely on the basis of "Darwin's ladder" is now followed by problematic consequences brought in its train.

Neo-Darwinism and Neo-Liberalism

In an interview with an Austrian daily paper[10] Richard Dawkins said, "No decent person wants to live in a society which works according to Darwinian laws. . . . A Darwinian society would be a Fascist state." Yet, what more humane society could exist if everything is evolution? Where is the freedom supposed to come from in order to oppose the "murderous Darwinian nightmare" (as Woody Allen described life)? The Viennese economist Ewald Walterskirchen points out the close relationship between neo-Darwinism and economic neo-liberalism:

> Both theories start from the belief that only random changes or adaptations determine the process of development, by way of selection, that is, of competition. The United States economist Paul Krugman is quite right when he writes that a textbook on neo-classical micro-economics reads like an introduction to microbiology.
>
> In economics, the close relation to biology is seen especially in the writings of Hayek, who is reckoned to be one of the fathers of neo-liberalism. Friedrich von Hayek, a member of a family of biologists, talks explicitly about a "sifting" by

[10] *Die Presse*, July 30, 2005, p. viii.

the market. Hayek holds that a high rate of unemployment—like an excess population in the animal world—is economically desirable, so that natural selection can have something to work on. In turn the OECD (Organization for Economic Cooperation and Development), the seat of neo-Liberalism, interprets the economic crisis in Europe simply as a lack of ability to adapt to shocks—just as neo-Darwinians interpret the dying out of animal species.

The conclusions drawn from these reflections in terms of economic policy are quite clear: economic policy only needs to create the right overall framework and conditions for the selective mechanism of the market to operate effectively. Decoded, this amounts to saying that we have to get rid of the European social model.[11]

Neo-Darwinism and the Pedagogy of Fitness

The main preoccupation in educational policy is the "economization" that has been more clearly observable for some years now. A fundamental paradigm of education today is adaptation, from the point of view of usefulness, especially as a preparation for the labor market. Key competencies like mobility and flexibility are high on the agenda, while the basic outline of Catholic social teaching, to the effect that the economy is there for man, and not vice versa, is being forgotten. Even that part of the fundamental task of education and school, to bring people up and educate them to resist certain things, is given little attention. We are living in peaceful times, but the question does arise: Who will be able clearly to articulate a well argued opposition if everyone is taught from an early age to trim their sails to the prevailing wind? No attention is paid to what

[11] *Der Standard*, July 16/17, 2005.

happens to those who cannot adapt—those who are too slow, not smart enough, not competitive enough, who are too reticent, too shy or unsure of themselves. They are pushed to one side. In the center stand the ones who have adapted, the useful ones—they are the elect, destined to survive. No thought is given, however, to the fact that it is precisely in periods of complete change that those best adapted to the conditions that have hitherto prevailed are the first to fall victim to change.

To resist ideological variants of evolutionism is one of the forms in which we may live in a free and responsible way today, even if there is a price to be paid. In his well-known lecture at the Sorbonne in Paris, Cardinal Ratzinger spoke about the fact that through its "choice in favor of the priority of reason" Christianity signifies today "enlightenment": that is, liberation from false dependencies.[12]

The Struggle for an Ethics of Life

Today, there is an intense ideological discussion with a materialistic view of the world in the great realm of bioethics. In broad areas, the Christian view of man, and in particular the teaching of the Catholic Church, is often alone nowadays in defending the unconditional dignity of man from conception to a natural death. Without being diverted by all the (frequently enormous) criticism, the Catholic Church insists that there is something like a "message from the Creator" in nature, and hence there is a morally binding order in creation that remains the guiding principle even in bioethics.

[12] Cf. J. Ratzinger, *Truth and Tolerance: Christianity and World Religions* (San Francisco: Ignatius Press, 2004), p. 181).

It is very strange: evolutionary ideology rejects any "designer" God. And yet projects abound in which *man* himself declares himself the "designer" of evolution. For instance, Simon Young quite openly advocates a biologistic, eugenic type of ethics.[13] Unsuccessful people are those who have not managed to "upgrade" themselves biologically.

In vitro fertilization, with all that follows from it, the "preimplantation diagnosis" that is to some extent already being practiced, the endless consequences of the "over-production" of embryos, with the discussion about their exploitation to obtain embryonic stem cells, for instance—all that leads to enormous problems, to breaking through ethical barriers, and, we must not forget, to a great deal of new human suffering, which is what people had claimed to be doing away with in the first place. The anxious objections of the Catholic Church are often left to die away unheeded.

The way that biotechnology itself is trying to take evolution in hand here inspires anything but trust. To justify all of these ventures of "bio-engineering" beyond the ethical limits, people always refer to the opportunites for positive interventions. Illnesses can be cured, people's desire for children can be fulfilled, goals of social improvement can be attained. Yet beware: a good many of these kinds of "progress" have turned out in the long run to involve problems. The enormous economic interests that are usually behind the biotechnical adventures often prevent us from being openly and honestly informed about the negative consequences of this "progress". One example of this are the results of research into the high risks involved in in vitro fertilization, which are hardly ever mentioned, even though they have been published.

[13] Simon Young, *Designer Evolution: A Transhumanist Manifesto* (New York: Prometheus, 2006); reviewed in *First Things* 164 (June–July, 2006): 48–49.

CONCLUSION

Gratitude and wonder, adoration and praise remain with us and grow. "O Lord my God, thou art very great!" (Ps 104:1). This wonder in our hearts must never die away. All the knowledge that I have been able to acquire over the course of years, even if it consists only of fragments of fragments of what can be known, has simply led me to marvel yet further—even about everything that can be learned from the exact science of evolution. The "Darwinian ladder" has made available to us—thanks also to genetics—a marvelous insight into the way life has ascended, the way it has come into existence and has been shaped and developed. The "Jacob's ladder" that connects this ascending and descending movement of life with heaven, with the activity of the living God, of his Logos and his creative Spirit, cannot replace the labors of research in "climbing" up the "Darwinian ladder". It does not tell us how the Creator made his works, how he has sustained them and guided their development. However, it does tell us with absolute certainty, more certainly than any scientific theories, that it is his Word, Christ, the Logos, through whom and toward whom everything has been created; and that his Spirit, who was moving over the face of the waters at the very beginning (Gen 1:2), and who is love, is moving in all created things and gives them meaning and purpose. Logos and agape, reason and love, are the material from which the world was made, of which it consists, and with which it is being perfected. It is well worth living in this conviction—and dying in it. For what kind of an evolution would it be if resurrection and eternal life were not its ultimate goal?

SCRIPTURE INDEX

INDEX OF PERSONS